AEI Legislative Analyses
Balanced analyses of current proposals before the Congress, prepared with the help of specialists in law, economics, and government

Natural Gas Proposals

1983
98th Congress
1st Session

American Enterprise Institute
for Public Policy Research
Washington and London

ISBN 0-8447-0249-8
Legislative Analysis No. 34, 98th Congress
September 1983

Contents

1

Introduction

The federal government imposed controls on producers' prices for natural gas sold in the interstate market following the mandate of a 1954 U.S. Supreme Court decision. Shortages of natural gas in the interstate market in the 1970s required the severe curtailment of consumers' use of natural gas. The Carter administration issued pessimistic predictions as to the future supply of this premier fuel and sought legislation to ameliorate the supply problem somewhat.

Legislation in 1978 addressed issues of gas supply and price by pursuing a variety of strategies. The interstate natural gas market, which had been experiencing artificially low prices, significant supply shortages, and increasing demand,[1] became one in which there were adequate supplies, rising prices, but declining demand. Policy experts describe the natural gas market as now being in total disarray.[2] Prices of perhaps 60 percent of the natural gas reaching the market will be decontrolled in 1985, thus creating additional uncertainties.

Policy makers are seeking new policy prescriptions to deal with natural gas supplies and prices. This legislative analysis will explore the thrust of past policies and their consequences. It will then examine the principal arguments for and against the thrust of the leading legislative proposals now being considered.

The reader should consult the glossary at the end of this publication for the meaning of terms employed in this analysis.

2

BACKGROUND

THE NATURAL GAS INDUSTRY

Natural gas accounts for approximately 27 percent of the energy consumed in the United States. It is used in approximately 55 percent of all residential and commercial establishments and provides approximately 40 percent of energy consumed by industry and agriculture.[3] The natural gas industry is composed of three major segments—production, transmission, and distribution—each of which is regulated by federal, state, and/or local authorities.

Producers. Producers explore for and develop natural gas reserves. Production companies that produce natural gas often do not provide for its transportation and subsequent sale for resale. As a result, these production companies are generally referred to as independent producers because they are unaffiliated with the transporting pipelines.[4] In addition to conducting exploration, drilling, and production operations, the producer may perform certain extraction operations. This is the process whereby usable liquid byproducts are removed from the gas before the gas is then placed into the purchasing pipeline's facilities.[5]

Natural gas production is subject to federal and state regulation. The price of natural gas produced at the wellhead is regulated under the Natural Gas Act of 1938 (NGA) and the Natural Gas Policy Act of 1978 (NGPA). State regulation generally affects such matters as the spacing of wells and the establishment of "allowable" production quantities to prevent drainage and protect correlative rights.[6]

Pipelines. This segment of the industry purchases, at the wellhead, natural gas from producers, and transports it to market for resale. At that point, the pipeline will resell the gas to local distribution companies for further resale. The nation's entire network of interstate and intrastate natural gas pipelines, including local distribution companies, extends over one million miles.[7]

Many pipeline companies, through affiliated companies, also produce their own natural gas.[8] They may also purchase gas from and sell to other pipelines. In November 1982, the major interstate pipeline companies produced 29 billion cubic feet (Bcf) of natural gas, purchased 801 Bcf from producers, imported 63 Bcf, and purchased 291 Bcf from others.[9] Pipelines also transport gas for customers who have obtained their own supply. In addition, pipelines may make direct sales of gas, usually to large industrial customers.

Distribution Companies. This segment of the industry consists primarily of local public utilities that purchase natural gas from pipelines. The distributors then resell the gas to end users, such as residential, commercial, or industrial customers.[10] There are approximately fifteen hundred local distribution companies operating in the United States that are regulated by municipal and state agencies.[11]

Pricing at the Wellhead

With the passage of the Natural Gas Act in 1938,[12] the Federal Power Commission and its successor agency, the Federal Energy Regulatory Commission,[13] have the authority to regulate the sale for resale of natural gas in interstate commerce.[14] Under the NGA the FERC has the authority to regulate the transportation of natural gas in interstate commerce.[15] Interstate natural gas pipelines are not common carriers under either the NGA or the NGPA.[16]

Until 1954, the FPC did not regulate the sale of natural gas at the wellhead. In that year, however, the U.S. Supreme Court in *Phillips Petroleum Co.* v. *Wisconsin,* held that independent natural gas producers are "natural gas compan[ies]" within the meaning of the Natural Gas Act.[17] Writing for the Court, Justice Sherman Minton stated:

> [W]e believe that the legislative history indicates a congressional intent to give the Commission jurisdiction over the rates of all wholesales of natural gas in interstate commerce, whether by a pipeline company or not and whether occurring before, during, or after transmission by an interstate pipeline company.[18]

During approximately the six years following *Phillips,* the FPC attempted to regulate the sales for resale of natural gas by producers on a company-by-company basis using a cost-of-service methodology.[19] This led to lengthy hearing procedures, however, and a large backlog in the number of cases. This period has been characterized as one of "lax approval of price increases" for producers.[20]

In 1960, the U.S. Supreme Court in its *CATCO* decision ordered the FPC to deny applications for new certificates of public convenience and necessity for sales of natural gas by producers "where the proposed price is . . . out of line or. . . might result in a triggering of general price rises or an increase in the applicant's existing rates."[21] Thereafter, the FPC began consideration of a new form of wellhead pricing, the application of "area rates" to gas from broad geographic producing areas of the country.[22]

In its first area rate proceedings, which involved the Permian Basin Area, the commission examined costs and set rates for the gas produced in that particular area.[23] The FPC also established a pricing structure for different vintages of gas. In order to stimulate exploration, for instance, the ceiling price for gas not yet under contract was established at a level higher than that for "old" gas.[24] The U.S. Supreme Court upheld the FPC's actions, concluding that the FPC had the authority to set industrywide rates, affirmed the FPC's pricing structure, and approved the "overall effect" of the area rate decision. The *Permian* case was the first of a number of court decisions reviewing the FPC's authority to establish area rates.[25]

In time, however, the FPC determined that area rate making was inadequate and, as a result, developed a procedure for establishing rates on a nationwide basis. In these proceedings, the FPC considered projections of future costs in setting rates.[26] Although the

FPC's national rate methodology was upheld as permissible under the NGA, the reviewing court of appeals found that the FPC had been "unsuccessful in generating additional natural gas reserves for the interstate market."[27]

In the early 1970s, shortages of natural gas supplies developed because of increased demand for this attractively priced, clean-burning fuel.[28] After a period of national debate during a time of severe natural gas shortages and curtailments, Congress adopted the NGPA.[29] The sponsors of the legislation believed that it would resolve the gas shortage and create a pricing structure which would provide adequate incentives for the development of existing and new natural gas supplies.[30]

The NGPA was an alternative to the total deregulation of natural gas sold at the wellhead.[31] It established a regulatory scheme that provided for gradual deregulation of most natural gas. The NGPA expanded federal regulation to include intrastate gas as well as gas sold in interstate commerce.[32] In addition, the NGPA established a methodology for calculating ceiling prices for numerous categories of natural gas. By regulating the price of gas sold at the wellhead to an intrastate purchaser, the NGPA eliminated many of the effects of the "dual market," including the advantage theretofore held by unregulated intrastate pipelines following *Phillips.*[33]

The NGPA established a method for the gradual phase-out of price controls on newly discovered natural gas as well as for certain intrastate gas. But gas produced from wells which had been drilled and were producing prior to 1977 would fall into a "forever regulated" status.[34] Price regulation of old gas under the NGPA would continue until those wells ceased production. The NGPA further established ceiling prices for the sale of old natural gas that would be adjusted periodically to reflect inflation.[35] For the principal categories of gas dealt with under the NGPA and current prices for these categories, see appendix A.

The NGPA also deregulated so-called high-cost gas under section 107 of the act. Included under this provision was gas produced from (1) levels below 15,000 feet from wells drilled after February 19, 1977; (2) geopressurized brine; (3) coal seams; and (4) Devonian shale. Section 107 gas accounted for approximately 7 percent of total U.S. production in 1982. Section 107(c)(5) of the NGPA also provides the FERC with the authority to implement incentive pricing for gas produced under conditions the commission determines to present "extraordinary risks or costs."[36]

The law provides that the price for gas from new reservoirs, from new onshore wells either 2.5 miles from or 1,000 feet deeper than the nearest pre-1977 well, and from new outer-continental-shelf (OCS) leases, are to be deregulated on January 1, 1985. Gas produced from wells which were nearer to producing wells, yet for which production is from below 5,000 feet, are to be deregulated on January 1, 1985. Natural gas produced from new onshore wells no deeper than 5,000 feet is scheduled to be decontrolled on July 1, 1987.

Contractual Clauses

The lawfulness of indefinite price escalators,[37] including area rate clauses,[38] has been subject to judicial review. After the *Phillips* decision, the FPC prohibited the operation of certain indefinite price escalator clauses which permitted upward price adjustments when a specified triggering event occurred.[39] Favored-nation clauses, for example, were prohibited by the FPC on the basis that the price that one producer receives for the sale of its gas has no bearing on the revenue requirements or costs of production of another producer.[40]

When the FPC later changed from area rate to national rate regulation, it permitted producers and pipelines to use area rate clauses to permit an escalation of the contract price to the national ceiling rate.[41] Under the NGPA, questions arose whether the area rate clauses would permit the escalation of the prices to be received under those contracts to the applicable NGPA price levels.[42]

Another form of contractual provision that has caused concern in the present pricing structure is the "take-or-pay" provision, which requires a purchaser under a contract to take certain quantities of natural gas or, if unable to take such quantities, nevertheless to pay the producer for those quantities. Some gas purchase contracts require interstate pipelines to take, or pay for, significant quantities of gas not taken.[43] Although most gas purchase contracts have make-up rights,[44] the take-or-pay contract provisions are argued to be responsible for significantly increasing costs passed on by interstate pipeline companies to consumers.[45]

Recovery of the Cost of Purchased Gas

In order for an interstate natural gas pipeline to recover its costs and a return on its investment, it must apply, pursuant to section 4 of the Natural Gas Act, to the FERC for changes in its rates.[46] A pipeline's recovery of its costs for purchased gas is pursuant to a different regulatory procedure, whereby the pipeline is permitted to adjust its rates, usually on a semiannual basis, to "track" changes in its purchased-gas costs.[47] Approval of such increases has been automatic. The NGPA further provides that pipelines may be guaranteed the recovery of their costs of purchased gas, absent fraud or abuse by the pipeline.[48]

Natural Gas Imports

Natural gas is imported into the United States from three countries: Canada, Mexico, and Algeria. Canadian natural gas is imported into this country at prices subject to approval by the Canadian National Energy Board and the U.S. government.[49] Gas from Mexico is sold to purchasers in the United States by Petroleos Mexicanos (Pemex), an agency of the Mexican government.[50] Gas imported from Algeria is in the form of liquefied natural gas (LNG). This gas is purchased by U.S. companies from Sonatrach, the Algerian national oil company.[51]

The authority and standards governing the regulation of imported natural gas are contained in section 3 of the NGA.[52] The regulatory duties under section 3 were originally the responsibility of the FPC but are now shared with the Department of Energy's Economic Regulatory Administration (ERA).[53] Primary jurisdiction to authorize imports of gas from foreign countries has been delegated to the administrator of the ERA, who must consider such factors as price, security of supply, national and regional needs for gas, and balance of payments.[54] The FERC exercises the other section 3 authority over these proposed imports or exports that has not been delegated to the ERA. The FERC has authority pursuant to sections 4, 5, and 7 of the Natural Gas Act to consider the construction and operation and the location of import facilities, as well as the authority to review resale and transportation rates whenever the import is to be resold or transported in interstate commerce.[55]

Under section 3 of the NGA, an import authorization may be issued unless it is determined that the importation "will not be consistent with the public interest."[56] Although the "public interest" standard under section 3 has been regarded as "substantially equivalent" to the test of "public convenience and necessity" under section 7 of the NGA,[57] there is a basic difference between the two standards. Whereas section 7 requires that the applicant demonstrate that its proposal will be in the public interest, section 3 shifts the burden to a finding by FERC that the proposed "importation will not be consistent with the public interest."[58]

A recent proceeding before the FERC suggests the possibility that the certificate issued under section 7 of the NGA for the sale or transportation of imported gas could be revoked or modified, provided there has been a violation of the terms of the certificate. In order to revoke a certificate issued under section 7 of the NGA, especially if substantial sums have been expanded and operations have begun, the FERC would require a "compelling showing of a fundamental shift of a long-term nature in the basic premises on which the certificate was issued."[59] The ERA has determined, however, that it can modify an import authorization under section 3 of the act, particularly if the price for the imported gas is no longer in the public interest.[60]

Transportation in Interstate Commerce

The NGA establishes federal jurisdiction over the transportation of natural gas in interstate commerce, and over the pipelines providing such transportation, as a part of a "comprehensive and effective regulatory scheme."[61] In addition, the U.S. Supreme Court has found that the scope of the FERC's authority to regulate pipelines and the transportation of gas has been extended to include curtailment of gas sold by interstate pipelines.[62] Neither the NGA nor the NGPA imposes common or contract carrier obligations on interstate natural gas pipelines.

Under section 311(a) of the Natural Gas Policy Act,[63] the FERC may permit an interstate pipeline to transport natural gas on behalf of any intrastate pipeline or local distribution company. The FERC also may authorize the transportation of gas by an intrastate pipeline for an interstate pipeline or local distribution company.[64] Transportation under section 311(a) of the act does not subject the transporting intrastate pipeline to the FERC's jurisdiction under the NGA. It also eliminates the requirement that an interstate pipeline obtain a certificate of public convenience and necessity pursuant to section 7(c) of the NGA.[65]

3

Legislative Proposals

As this is written, the following significant legislative proposals have been introduced in the 98th Congress. Each of the proposals, which are discussed in detail in this chapter, is summarized and identified with its principal sponsors.

The Administration's Bill, "Natural Gas Consumer Regulatory Reform Amendments of 1983" (S. 615: Sen. James A. McClure, R-Idaho; H.R. 1760: Rep. Thomas Corcoran, R-Ill. and six cosponsors). The administration's bill deregulates the sale of natural gas at the wellhead. It provides for total deregulation of all gas by January 1, 1986. It would permit the immediate decontrol of natural gas sold under new contracts and under contracts that are renegotiated after the date of its enactment. It also would remove price controls for the sale of old gas. Gas due to be decontrolled under the NGPA in 1985 would remain under price regulation until January 1, 1986, with the price generally limited by a "price cap" based on the average price of decontrolled new and renegotiated contracts.

The price cap under the administration's bill is a volume-weighted average price calculated monthly on estimates of the prices to be paid and the volumes to be delivered during the first three months under new or renegotiated contracts. Beginning in the fourth month of the bill's enactment, the price cap would be recalculated monthly. These price limits will terminate on January 1, 1986, or earlier if a new or renegotiated contract is entered into.

Under the administration's bill, natural gas, other than high-cost natural gas, which would remain regulated under the NGPA, would be subject to a ceiling price of either the lower of the applicable NGPA price or the price cap. High-cost tight-formation gas (see the glossary) will be limited to the NGPA price in effect on the date of enactment of the administration's bill subject to the FERC's existing authority to lower the incentive price for this gas. S. 615 further provides that other high-cost gas which has not been deregulated will be priced at the higher of the contract price on the date of enactment or the price cap.

The NGPA currently permits the reimposition of price controls for an eighteen-month period by either the president or Congress.[66] The administration's bill would eliminate the possibility of the president or Congress reimposing price controls.

Most of the legislative proposals would restrict the operation of take-or-pay clauses in an attempt to reduce payments for new and high-cost supplies of gas. As for natural gas contracts in effect on the day of its enactment, S. 615 provides that a pipeline purchaser has the right to refuse deliveries of gas which exceed 70 percent of available deliverability, without incurring any obligation for the volumes not taken. This provision would apply only to deliveries under a contract from the date of enactment until December 31, 1985.[67] Casinghead natural gas is not covered by this provision. The FERC could limit the application of this provision to prevent drainage and protect correlative rights.

The bill provides that either party to a gas purchase contract in effect on the date of its enactment, and which has not thereafter been amended, may terminate such contract beginning January 1, 1985. The pipeline would have an obligation to transport natural gas for a producer that was party to the terminated contract at the rate of five cents per million Btu (MMBtu), unless the FERC establishes a different rate, plus the cost of transportation.

The administration's bill, as well as much of the other proposed legislation, would limit the amount of purchased-gas costs to be passed through by an interstate pipeline company. In essence, S. 615 provides for a "cap" on purchased-gas pass-throughs. It provides that between the date of enactment and December 31, 1985, there would be no automatic pass-through of purchased-gas costs above the pipeline's average gas costs in the month before enactment plus an allowance for inflation. Any increases above that level would require prior approval by the FERC before they could become effective. The administration's proposal would provide for a special FERC proceeding to consider purchased-gas costs. Such costs would not, as is now the practice, be put into effect, subject to refund, pending the outcome of the proceeding. The bill would not revise section 110 of the NGPA that permits the recovery of certain production-related costs.

The administration's bill also provides for a limitation on the pass-through of costs associated with the purchase of gas from an affiliate of the interstate pipeline. This proposal limits the pipeline's recovery of costs associated with its own production or purchased from an affiliate.

Incremental pricing under Title II of the NGPA was established to protect high-priority natural gas users, such as householders, from the expected price increases under the wellhead pricing provisions of the NGPA.[68] The FERC implemented incremental pricing regulations which required certain industrial users to bear the incremental costs of costlier gas purchases until the price to those users rose to the level of prices for alternative fuels. The administration's bill would eliminate the incremental pricing requirements established by Title II of the NGPA.

Interstate natural gas pipelines generally operate as buyers and sellers of gas. They also transport gas and to a lesser extent serve as private contract carriers. Essentially, under present law, no common carriage obligations are imposed on interstate natural gas suppliers.

Currently the FERC cannot, except in limited circumstances provided in other statutes, compel pipelines to carry gas for other parties. Approximately a dozen bills,[69] including the administration's bill, have proposed altering the status of natural gas pipelines to either common or contract carriers. Title IV of the administration's bill would give the FERC the authority to order any interstate pipeline to transport natural gas on behalf of a producer or a purchaser on a contract basis. Section 403 of S. 615 would add the following new section 317 to the NGPA:

> *Sec. 317. Contract Carrier Authorization*
>
> (a) In General—Upon application by a producer of natural gas or by a purchaser of natural gas from a producer, the Commission shall order any interstate pipeline to carry gas under contract between producer and purchaser upon such terms and subject to such conditions as it considers just and reasonable if the Commission finds that such pipeline has available capacity, that no undue burden will be placed upon such pipeline, that no construction

of new facilities would be required, and that such order would not impair the ability of such pipeline to render adequate service to its existing customers. The Commission may implement this section by rule or order.

In addition, the administration's bill would provide compensation for any transportation provided under this section of five cents per MMBtu profit for the pipeline in addition to its costs.

The administration's bill allows intrastate pipelines and local distribution companies to gain access to gas that they cannot obtain under current law. The administration's bill would exempt from the FERC's jurisdiction their acquisition of: (1) gas acquired in a first sale which, under section 601(a)(1)(A) and (B) of the NGPA, is not subject to the jurisdiction of the commission under the NGA; (2) the first sale of gas produced on the outer continental shelf; (3) a sale or assignment of gas under sections 302(a), 311(b), 312(a) of the NGPA; (4) a first sale under the "contract carrier" provision of the administration's bill; and (5) a sale for resale by a Hinshaw pipeline. This access provision would apply only to natural gas which remains in the state where the purchaser in the covered transaction receives the gas. In addition, the administration's bill would amend the Natural Gas Policy Act to provide a "nondiscrimination provision" to prevent the FERC from discriminating against either intrastate or interstate purchasers of gas in its exercise of jurisdiction over any gas pursuant to the NGA or NGPA.

At the present time, NGPA section 311(b) provides the FERC with the authority to permit an intrastate pipeline to sell natural gas to: (1) any interstate pipeline; or (2) any local distribution company which is served by an interstate pipeline. The administration's bill would expand the FERC's authority to allow the commission, by rule or order, to authorize any pipeline or local distribution company to sell natural gas to any other pipeline or local distribution company.

The administration's bill would expand the authority under NGPA section 312 for the assignment of natural gas. At the present time, the FERC may authorize any intrastate pipeline to assign to any interstate pipeline or local distribution company all or any portion of the intrastate pipeline's right to receive surplus natural gas under any first sale. The administration's bill would expand that authority to permit the commission to authorize any pipeline or local distribution company to assign surplus gas to any other pipeline or local distribution company. The administration's bill would broaden the definition of "surplus natural gas" to include any natural gas that exceeds the then current demands of the pipeline or local distribution company for natural gas as determined by the FERC or the appropriate state regulatory body.

The administration's bill does not address the question of imported natural gas or LNG.

The Senate Committee Bill, S. 1715 (Sen. McClure). The Senate committee bill, reported without recommendation July 29, 1983, was developed through the markup of a committee print authored by Senators McClure, J. Bennett Johnston (D-La.), Wendell H. Ford (D-Ky.). S. 1715 would remove wellhead price controls upon enactment of the bill for gas subject to first-sale contracts executed thereafter, and for contracts renegotiated thereafter if the renegotiated contracts expressly provide that controls will not apply. The bill would also decontrol gas released from take-or-pay requirements by the bill.

S. 1715 would, generally speaking, remove any remaining wellhead price controls the forty-fourth month after enactment of the bill. It would generally decontrol high-priced gas as defined in the bill the first day of the twentieth month after enactment. The bill would repeal provisions of the NGPA allowing the president or the Congress to reimpose wellhead price controls.

The FERC could deny a purchased-gas adjustment during the eighth through the forty-fourth months after enactment of the bill if the price of the gas did not meet a prudence test. Purchases would be deemed prudent if the weighted average amount paid during any month under a new or renegotiated contract did not exceed 110 percent of the free market indicator established by the bill. Purchases would also be deemed prudent if made in the exercise of the purchaser's right under a terminated contract to match the best offer obtained by the seller for gas under that contract or the amount paid is pursuant to a first offer by the purchaser for market-out gas or for certain other gas designated in the bill when the seller has terminated the contract.

The FERC could find that the cost of purchased gas exceeding 110 percent of the free market index to be prudently incurred under certain circumstances. It would make this judgment by considering the necessity for incurring these costs to acquire sufficient gas supplies to ensure adequate service to existing customers and, also, by considering the availability of lower-cost supplies. Interstate pipelines could not pass through the costs of gas purchased from an affiliate on contracts renegotiated after the date of the bill's enactment if the terms were not commensurate with those of contracts renegotiated with nonaffiliated producers. The bill would require the FERC to issue rules compelling pipelines to pass through the benefits of reduced purchased-gas costs, the reduced costs from decreases in take-or-pay volumes, and price reductions realized from applying the transitional price adjustment mechanism established by the bill.

The FERC could authorize any pipeline or local distributing company (LDC) to transport gas for any person, or sell gas or assign surplus gas to any other pipeline or LDC. The bill would broaden the authority of intrastate pipelines and LDCs to obtain gas from interstate sources for use solely within the state, without subjecting themselves to FERC jurisdiction. Interstate pipelines could not sell to an affiliated intrastate pipeline, Hinshaw pipeline, or LDC, a greater percentage of gas available to it under contracts than the percentage of available lower-priced gas which the affiliate is purchasing during the same month from nonaffiliated interstate pipelines.

S. 1715 would repeal certain restrictions on the use of gas imposed by the Fuel Use Act. It would also repeal the incremental pricing provisions of the NGPA. The bill would prohibit the sale of excess natural gas to customers outside the pipeline's historic market area at a price less than the applicable price indicator plus the just and reasonable rate for transportation. This provision would sunset in forty-four months. The bill would provide a limited antitrust immunity to independent producer cooperatives marketing gas released under market-out rights conferred by the bill or released from take-or-pay requirements.

The transitional provisions of the bill would permit purchasers of natural gas to reduce take-or-pay requirements during the first three years following enactment of the bill. Reductions would be allowed for a period of one year, a period that could be extended for an additional year but not beyond the end of the three-year period. Take-or-pay requirements could be reduced to 50 percent of deliverability during the first year after enactment of the bill, to 60 percent during the second year, and to 70 percent during the third year. The right to make this reduction would not extend to casinghead gas, to the volume of gas for which delivery is required to prevent drainage, or to the extent reduced volume might damage the well or reservoir.

The bill would allow either party to a gas sales contract in effect when the bill is enacted to market out or terminate the contract on sixty days' notice. Market-out authority would commence the twentieth month after enactment, in the case of high-priced gas, and with the forty-fourth month in the case of low-priced gas as defined in the bill. If either party terminates a contract, the other party could compel continued performance at a price equal to the free market indicator for a period of time as provided in the bill. S. 1715 contains quite detailed provisions allowing a purchaser under a contract that is terminated under market-out authority or per the price adjustment provisions of the bill to continue receiving the volumes of gas called for by the terminated contract. The former purchaser would have a right of first refusal permitting it to match the terms of an offer by a replacement purchaser for the volumes called for by the terminated contract. A first-offer right would permit the former purchaser to initiate the bidding process for the released gas. Although the seller could dispose of the gas to another purchaser, it could only do so if the other bidder offered a better price than that offered by the former purchaser.

S. 1715 contains elaborate provisions for gradually adjusting the price of gas under designated contracts either downward, as in the case of high-priced gas, or upward, as in the case of old gas, during the transition to a free market price and the ending of controls. A temporary price indicator would be established initially, and this would be replaced on the first day of the eighth full month after enactment by a free market indicator. This indicator would be used as the base point of reference in the phase-up and phase-down of gas prices. The indicator would be calculated monthly from weighted averages of gas sold under contracts of sale executed after enactment of the bill at decontrolled prices, and from contract prices for imports authorized during the six months prior to the monthly index.

The phase-up and phase-down transition pricing provisions would apply only to contracts for gas produced in the contiguous forty-eight states and adjacent outer continental shelf and only if the buyer or seller elects to come within the pricing provisions. The provisions would not apply to contracts whose prices would be deregulated as stated in the first paragraph of this description. The election to come under the transition pricing provisions would have to be made by the first day of the fifth month following enactment of the bill. If one party elected to come under these pricing provisions, the other party could agree or it could terminate the contract. A purchaser who first agrees to the seller's proposal to apply these pricing provisions to gas selling below the price indicator could change that election at the end of the eighteenth month of the thirty-six-month adjustment period. Performance under terminated contracts would be required under the circumstances cited in the eighth paragraph of this description for a period of time that would depend on the ability of the seller to find a new purchaser, the categories of gas represented, and the provisions of section 131 of the NGPA as added by the bill.

Section 104 of the bill would require a pipeline that was a purchaser under a contract terminated either under the market-out or transitional pricing provisions of the bill to provide contract carriage for gas under the terminated contract. In addition, a pipeline reducing quantities of gas taken according to the bill's take-or-pay provisions would have to provide carriage for the gas so released from the contract. Transportation charges for gas carried pursuant to section 104 would include an incentive allowance of five cents per MMBtu in addition to the regulated rate for that transportation.

Section 401 of the bill, derived from a proposal by Senator Bill Bradley (D-N.J.), attempts to encourage increased competition in the natural gas industry by requiring contract carriage of natural gas. The bill requires gas pipelines and LDCs to transport gas for a buyer or seller on request if they have the capacity. Intrastate pipelines and LDCs would be exempt from FERC jurisdiction if state officials certify they have appropriate authority to compel contract carriage. The bill establishes a rebuttable presumption that pipelines and LDCs have available capacity. The presumption could be rebutted by convincing the FERC that total capacity is required to render adequate service to existing customers. The bill establishes priorities for service and provisions for dealing with curtailments. Pipelines and LDCs could be ordered to construct minor facilities necessary for the contract carriage of gas, at the expense of the person requesting construction. The incentive compensation rate cited above would apply to voluntary contract carriage until another rate is established by the FERC as more appropriate. The incentive rate would not apply to carriage performed for an affiliate. Pipelines and LDCs protesting an application for contract carriage would be allowed an incentive allowance at the FERC's discretion not exceeding five cents per MMBtu if the commission finds that the protest was a good-faith effort to protect the pipeline's or the LDC's ability to continue to render adequate service to existing customers.

Clean Bill of the House Subcommittee on Fossil and Synthetic Fuels (July 29, 1983). The Subcommittee on Fossil and Synthetic Fuels considered a committee print offered by the subcommittee chairman, Representative Philip Sharp (D-Ind.). The subcommittee reported that bill as substantially modified by an amendment offered by Representatives Richard C. Shelby (D-Ala.) and Tom Corcoran (R-Ill.) as a clean bill for consideration by the House Energy and Commerce Committee.

Under this bill, gas from wells commenced after the enactment of the bill would not be regulated. Newly enhanced flow of gas from existing wells would be decontrolled. Gas from onshore wells whose production is under sales contract on the date of the bill's enactment would not be controlled if the wells are located two and a half miles from the nearest marker well, as defined in the NGPA, or are completed to a depth of 1,000 feet below that of each marker well within two and a half miles, so-called infill drilling incentive wells. Gas produced to prolong the productive life of a well would be excluded from price controls if the purchaser agrees, but this exception would not apply in the case of gas produced by an affiliate of the purchaser. Gas released from delivery under a gas sales contract by provisions of the bill dealing with take-or-pay clauses, force majeuer (roughly the equivalent of an act of God), or other rule of law or contract provision permitting the purchaser to take or pay for 50 percent or less of the volume contracted for would be decontrolled.

Other gas would remain controlled or would be decontrolled on the schedule provided in the NGPA subject to an exception. Natural gas that would be decontrolled under section 121 of the NGPA as amended by the bill and which is subject to an indefinite price escalator clause or price redetermination clause could not be sold for more than the amount set by a natural gas price indicator. That indicator would be based on a volume-weighted average of gas delivered to interstate pipelines over a designated three-month period. The provisions of the NGPA permitting the reimposition of price controls would be repealed.

Purchased-gas adjustments would be subjected to FERC hearings on the protest of any interested party. Misrepresentation, imprudence of the pipeline, its failure to bargain at arm's length with any producer, and operating under a contract that prevents responses to changes in customer demand or other market forces would, on the FERC's determination, be "abuses" that would justify denial of the automatic pass-through of purchased-gas costs.

The FERC could not approve any rate or charge unless the rate design places the pipeline at risk that it will not collect its allowed return on equity if its purchasing and operating practices do not "meet the demands of its market and prevent curtailment."

Take-or-pay requirements in gas sales contracts could be limited to 50 percent of the contracted volume for three years after enactment of the bill, at the purchaser's election. This provision would not apply to casinghead gas or occluded natural gas from coal seams. The FERC could exclude from this provision volumes of gas necessary to prevent drainage or to protect correlative rights. If a pipeline's obligation to take gas is reduced under this provision of the bill, it must, under rules to be established by the FERC, reduce the amount of gas that its customer is required to take. Purchasers from an interstate pipeline that are required by contract or tariff to take specified quantities of gas could, from the sixth through the seventeenth months after enactment of the bill, elect not to take or pay for any portion of the gas. Contract clauses requiring these purchasers to take minimum quantities of gas would be unenforceable if these clauses did not permit make-up within a time specified by the FERC.

A pipeline refusing to take gas in excess of 50 percent of its contractual take-or-pay obligation would give the seller thirty days' written notice of refusal to take that volume. The seller could terminate the contract as to the amounts not taken and negotiate for the resale of that gas. The original contract purchaser, however, would have a right to take the gas on the terms negotiated for the resale of the gas. A producer of newly enhanced gas and infill drilling incentive well gas could give the purchaser of the producer's other gas the opportunity to buy this additional gas. If the purchaser refused the gas, the producer would be relieved of contractual and legal obligations to the purchaser. The purchaser would have the right to purchase that gas on the terms and conditions worked out with another purchaser for a fifteen-day period.

The bill contains extensive provisions dealing with required contract carriage by pipelines. Pipelines electing not to take gas under take-or-pay clauses, or refusing to purchase new enhancement gas or gas from infill drilling incentive wells, would be required to carry that gas for the producer to the extent of available capacity. Pipelines would be required to carry gas on a firm or interruptible basis for other shippers without discrimination if these shippers agree to pay applicable tariff rates. Intrastate pipelines subject to state regulation that requires carriage of gas on a nondiscriminatory basis would be exempt from federal contract carriage requirements. The bill provides for allocation of pipeline capacity and priority of service. The FERC would issue regulations to determine if available capacity for contract carriage exists. A pipeline could be required to build and operate minor facilities at the shipper's cost. Local distribution companies would not be subject to the contract carriage provisions.

Various provisions of the Fuel Use Act would be repealed. The bill would also repeal

Title II of the NGPA requiring incremental pricing.

The bill would permit intrastate pipelines and local distributing companies access to gas supplies that they cannot tap under current law. The FERC would be prohibited from discriminating against either intrastate or interstate purchasers of natural gas. Interstate pipelines could not discriminate in the allocation of their lower-priced gas in favor of affiliated intrastate pipelines, Hinshaw pipelines, or local distributing companies. The FERC could authorize a pipeline or local distribution company to assign gas surplus to current demands to another pipeline or distributing company.

Orders for the importation of natural gas and LNG signed before the effective date of the legislation could be revoked beginning 270 days after enactment until the import contracts were renegotiated at prices responsive to the then current market for natural gas and responsive to periodic changes in demand for gas. Contracts for an importer all of whose gas sales are of imported gas would be excluded from this provision of the bill, thus exempting the Distrigas contract in Boston. Prudently incurred costs of constructing facilities for handling Algerian gas would, in the event of revocation, be recoverable by the importer, as in the case of the Panhandle/Trunkline facility at Lake Charles, Louisiana. The FERC would, however, be required to deny a rate of return on such costs.

The Consumer Relief Bill, "Natural Gas Consumer Relief Act" (S. 996: Sen. Nancy L. Kassebaum, R-Kans.; H.R. 2154, Rep. Richard A. Gephardt, D-Mo., and seventy-three cosponsors). The Consumer Relief bill does not provide for wellhead price decontrol but extends NGPA price controls for two years. It would roll back prices to past levels for nearly every category of gas under the NGPA. The House version rolls prices back to January 1982 levels, while its Senate counterpart rolls prices back to August 1982 with limited adjustments for inflation. The Consumer Relief bills also place a ceiling price on previously deregulated high-cost gas and lower the existing ceiling price for gas produced from tight formations.

In contrast to the administration's bill, old gas would remain "forever regulated" under the Consumer Relief bills. Gas which is to be deregulated under the NGPA in 1985 and 1987 would be regulated for an additional two-year period. The FERC's authority to increase certain prices of certain categories of gas, including old gas, under the NGPA would be repealed by the Consumer Relief bills, as would the commission's authority to add new categories of high-cost gas to those eligible for incentive pricing under NGPA section 107(c)(5).

The two Consumer Relief bills would, in effect, reimpose price controls on high-cost gas and provide limitations on tight-formation gas. The House bill provides a maximum lawful price for high-cost (non-tight-formation) gas at 150 percent of the price for NGPA section 103 gas (relating to new onshore production wells). The FERC would be given the authority to prescribe a maximum lawful price to provide "reasonable incentives" to produce such gas. The price determined by the commission, however, could not exceed the ceiling of 150 percent of the NGPA section 103 level. The Senate version of the Consumer Relief bill would permit the FERC to prescribe a maximum lawful price in excess of the applicable price level for all high-cost gas, including tight sands. That increased rate would have to be "just and reasonable" under the NGA and reflect what is "necessary to allow the producer to earn a just and reasonable return," taking into account any cost the producer recovers as a result of section 110(b) of the NGPA.[70]

The Consumer Relief bills differ from the administration's bill with regard to the monthly escalation of ceiling prices. A new modified price adjustment factor would be applied to January 1982 NGPA prices (in the case of S. 996, it would be applied to August 1982 NGPA prices) to determine the ceiling price for the first month after the enactment of the bill, and each succeeding month thereafter. Specifically, the Consumer Relief bills would adjust the ceiling prices to reflect only 75 percent of the monthly change in the GNP implicit price deflator or the Department of Labor's energy index. The index that produces the least increase, or largest decrease, in the monthly adjustment would be used to calculate the inflation adjustment in the price of natural gas.

The Consumer Relief bills would permit the FERC to allow a natural gas producer to collect a price higher than the new ceiling price if: (1) the well was drilled on or after January 1, 1982 (or August 1, 1982, in the case of S. 996), and before enactment of the bill, and (2) the higher price is just and reasonable under the NGA. The price approved by the FERC, however, may be no higher than the applicable NGPA price for the date surface drilling on the well began, as adjusted by the modified price adjustment factor provided in the bill.

The administration's bill would not revise NGPA section 110, which permits the recovery of certain production-related costs. The Consumer Relief bills would, however, amend section 110 of the NGPA to provide that the maximum lawful price would include the recovery of all costs and profits associated with the production and delivery of marketable natural gas.

These bills would provide that take-or-pay clauses in excess of 50 percent of the maximum annual volume the pipeline has contracted to take will be unenforceable. This would apply for the sale of gas from new wells (NGPA section 102), new onshore production wells (NGPA section 103), high-cost gas wells (NGPA section 107), and stripper wells (NGPA section 108). In addition, contracts which do not contain a one-year make-up provision would be unenforceable.

The Consumer Relief bills would declare indefinite price escalator clauses to be against public policy and unenforceable. This legislation would void any contractual provision that provides for the establishment or adjustment of the price for gas delivered by reference to prices for natural gas, crude oil, refined petroleum products, or any other commodity.

The Consumer Relief bills would authorize an interstate or intrastate pipeline to notify its seller that it wishes to renegotiate the terms of its contract. If unsuccessful in the renegotiations, the pipeline would be able to market out. Unlike the administration's bill, the seller would not be given the authority to market out. Should the pipeline determine, in its sole discretion, that it would be unable to market gas that is required to be taken under the contract, it may market out by refusing to accept delivery of any or all volumes of gas under the contract without incurring an obligation to pay any fee or obligation with respect to such volumes.

The pipeline must, however, meet the following conditions: (1) the pipeline must reduce its highest-cost gas first; (2) the pipeline must reduce delivery of equally priced gas from affiliates before reducing its takes of gas from nonaffiliated producers; and (3) producers who have been marketed out by the pipeline are to be given the right to sell their gas to the pipeline at a price equivalent to that offered by any other producer prior to the pipeline entering into a new contract.

The Consumer Relief bills would prevent a pipeline that has exercised its market-out authority from accepting, during the period of reduction of any delivery, other volumes of gas at a price equal to or higher than the gas from the original seller, if the deliveries are made under a new or amended contract entered into after the exercise of the market-out authority. A pipeline that has exercised its market-out authority would have a right of first refusal to purchase gas which the producer has renegotiated to sell to another purchaser. If gas which the original pipeline has refused to purchase is then sold to another purchaser, the pipeline exercising market-out rights must provide transportation of such gas.

The Consumer Relief bills would expand the FERC's authority to deny the pass-through of purchased gas costs by a pipeline. In addition to the "fraud, abuse or similar grounds" standard under section 601(c)(2) of the NGPA, the House bill would permit the FERC to deny pass-through if it determines that the pipeline "has failed to adopt and follow practices which minimize amounts paid for purchases of natural gas." The Senate bill seeks to strengthen NGPA section 601(c) by adding "imprudence" to the "fraud, abuse or other similar grounds" test.

This legislation would require the FERC to complete, within one year, a rule making to ensure that prices paid by interstate pipelines for gas purchases are "sensitive to end-use market conditions" and that "end-use market signals are effectively transmitted to pipelines and producers." The commission also is to review the tariffs of interstate pipelines to ensure that the standards are met, and it can require tariff modifications from those that fail to meet the standards.

The Consumer Relief bills would make unenforceable: (1) pipeline minimum quantity bill requirements in excess of 50 percent of the maximum annual volume which the purchaser has contracted to receive, or (2) minimum bill requirements which do not permit the purchaser to take delivery of gas after the date of payment. The provision would be applicable to the sale of gas by other than an interstate or intrastate pipeline.

The bills authorize the FERC to require pipelines to carry gas for producers, but such carriage must not impair a pipeline's ability to serve existing customers or subsequently acquired customers who are entitled to uninterruptible service. The bills address the issue of exempting intrastate pipelines and local distribution companies from the FERC's jurisdiction in conjunction with these contract carriage provisions. The bills provide that intrastate pipelines or local distribution companies do not submit themselves to the FERC's jurisdiction under the NGA merely because such companies have exercised their rights to secure transportation of natural gas under the new contract carrier authorization. The bills further provide that any transportation or sale of natural gas that takes place beyond the point at which the intrastate pipeline or local distribution company receives the gas into its facilities shall not be considered in interstate commerce within the meaning of the NGA to the extent that such natural gas remains in the state of purchase.

The Consumer Relief bills would amend section 3 of the NGA to provide a price cap on certain imported natural gas. These bills propose that natural gas imported from any foreign country, other than Mexico or Canada, may not be imported if the first sale price of the gas within the United States exceeds 150 percent of the maximum lawful price under section 103 of the NGPA for the month of importation. The Consumer Relief bills also require the president to submit a report to the Congress within sixty days of enactment on the status of negotiations with the Canadian government regarding modification of the border price for gas imported into the United States from Canada.

The Slattery-Coats Bill, "Natural Gas Policy Act Reform Amendments of 1983" (H.R. 2508: Reps. Jim Slattery, D-Kans., and Daniel R. Coats, R-Ind.). This bill would continue price controls on old gas, give either party to a gas purchase contract certain market-out rights, and limit take-or-pay clauses to 50 percent of deliverability. Natural gas discovered after the bill's enactment would be decontrolled, and gas discovered since April 20, 1977, may be decontrolled upon contract renegotiation. A prudence standard would apply to the purchasing practices of pipelines. The Fuel Use Act provisions restricting the use of gas by industrial users would be repealed. In addition, the secretary of energy, in consultation with the secretary of state, would report to Congress on the status of natural gas imports.

The Slattery-Coats bill would eliminate price controls on natural gas produced from wells: (1) which are drilled on or after the date of its enactment, (2) which are subject to a contract entered into on or after the date of enactment, and (3) which were not dedicated to interstate commerce on April 20, 1977. The prices for certain other categories of regulated gas could be decontrolled if a renegotiated or amended contract so provides. The sale of gas under a new contract between the producer and the original purchaser or an unaffiliated purchaser resulting from the exercise of the market-out authority under the bill would not be subject to price regulation if the parties so agreed.

This legislation would not alter the ceiling price methodology existing under current law. The bill would also eliminate the possibility of the president or Congress reimposing price controls.

The Slattery-Coats bill would make a take-or-pay provision unenforceable to the extent that it fails to provide adequately for: (1) make-up of all gas volumes that are paid for but not taken; and (2) if make-up does not take place, refunds of all amounts paid for gas quantities which are not taken. For the three-year period from the date of enactment, contract terms which impose take-or-pay requirements in excess of 50 percent of deliverability would be declared against public policy and unenforceable except for casinghead and stripper well gas.[71] The FERC may permit take-or-pay provisions in excess of 50 percent to the extent necessary to prevent drainage or reservoir damage or other severe and irreparable financial injury to the producer. The legislation also provides that any favored-nation provision in any contract for the first sale of natural gas would be unenforceable.

The Slattery-Coats bill further provides that both the producer and the pipeline have the right to market out. Specifically, the bill states that for one year after the date of enactment either party to the contract may, in its sole discretion and after notice to the other party, terminate any purchase contract, other than a warranty contract, for the first sale of new natural gas, gas from the outer continental shelf, gas from new onshore wells which were not dedicated to interstate commerce on April 20, 1977, and high-cost gas. This market-out authority may be used only once for each contract, and, once it is used, the seller may either negotiate a new or amended contract with the original purchaser or obtain a new purchaser. Moreover, if a contract has been negotiated with a new purchaser, the producer must offer to the original purchaser a right of first refusal at the same price and at reasonably comparable terms to those negotiated with the new purchaser. Should the original purchaser decline, then it would have the obligation to transport the natural gas on which it has exercised its market-out right.

The Slattery-Coats bill would eliminate "abuse" as a basis to deny pass-through of purchased-gas costs by a pipeline. Instead it would adopt a prudence standard, to be defined by an FERC rule making. Should the commission determine that any purchase violates the standard, it may deny the pipeline the recovery of such costs paid under a contract that is entered into or renegotiated after the effective date of the establishment of its prudence rule.

The bill would amend the NGPA to require the commission, as expeditiously as practical, to issue a general statement of policy respecting rate design principles for application to individual interstate pipelines on a case-by-case basis. This proposal requires that the commission's statement "ensure that jurisdictional pipeline rates are sensitive to end-use market, supply-and-demand conditions." The legislation also provides that the FERC must adjust contracts between pipelines, and between pipelines and distributors, to the extent necessary to make minimum-bill provisions correspond to any reduced volumes taken as a result of application of take-or-pay limitations. The Slattery-Coats bill would eliminate the incremental pricing requirements established by Title II of the NGPA.

The Slattery-Coats bill would require the submission to Congress of a report concerning natural gas imports. The secretary of energy, in consultation with the secretary of state, would be required to review the exercise of regulatory authority over natural gas imports and recommend to the Congress action, if any, which is necessary to ensure that the contractual terms and other requirements are in the public interest. The bill would direct the secretary of energy to take into consideration: (1) the foreign policy and foreign trade objectives of the United States; (2) the long-term gas supply needs of the United States; and (3) the findings, purposes, and requirements of the NGPA.

The Heinz Bill, "Natural Gas Policy Act Amendments of 1983" (S. 689: Sen. John Heinz, R-Pa.; H.R. 1752: Rep. Joseph P. Addabbo, D-N.Y.). This bill would not change the schedule for the decontrol of gas established under the NGPA. The NGPA inflation adjustment factor would be repealed. Either party to a gas purchase contract would have market-out rights. The pipeline buyer of market-out gas has a right of first refusal over that gas but an obligation to provide transportation, on a best-efforts basis, if it does not purchase the gas. Take-or-pay clauses are limited to 50 percent of deliverability, and indefinite price escalators are nullified. Pipeline purchasing practices are to be reviewed under the "just and reasonable" standard of the NGA. The FERC is to examine the operation of minimum bill provisions in the tariffs of interstate pipelines.

The Heinz bill addresses the treatment of value received by the producer for the transportation of gas by pipelines which contain nongaseous hydrocarbons ("liquids and liquefiables"). It would require the FERC to reduce ceiling prices by the value of any nonmonetary consideration[72] received by the seller in addition to the maximum lawful price.

The Heinz bill would also establish the appropriate Btu measurement determination to be made by the FERC in pricing gas. Before the NGPA, the ceiling price for natural gas sold in interstate commerce was adjusted for Btu content that was calculated to account for certain standard conditions, one being the saturation of gas with water vapor. After the act became law, the FERC announced a change in natural gas pricing, such that Btu content formerly measured in standard conditions would be measured on an "as delivered" or actual water content basis. The use of an "as delivered" Btu measurement is believed to result in a higher unit price for gas sold by producers than the use of the "saturated with water vapor" method. The Heinz bill would require the Btu content of natural gas deliveries to be measured for purposes of applying maximum lawful prices on a "saturated with water vapor" basis, unless some other basis is expressly provided in the contract. Thus, the Btu measurement would revert to the earlier standard applied by the FPC.

The Heinz bill would declare indefinite price escalator clauses in gas purchase contracts to be null and void. It would limit the operation of area rate clauses.

The bill also contains provisions allowing either party to alter the terms of the gas purchase contract. All contracts for the sale of new natural gas, onshore gas which was not committed to interstate commerce on April 20, 1977, and high-cost gas are deemed to include market-out provisions. The Heinz bill also would require the purchaser to provide transportation on a best-efforts basis if it refuses to purchase the gas under the market-out provisions of the bill and the seller obtains another buyer.

The Heinz bill would require the FERC to complete within one year a study of minimum bill and related provisions in the tariffs of interstate pipeline companies. The study would analyze such matters as the responsiveness of wellhead prices to retail market conditions, the rates charged customers of interstate pipelines, the ability of local distribution companies to secure varied sources of gas supply, and the present and future financing of existing and new interstate pipelines. The FERC could require the revision of minimum bill requirements in the interim.

The Heinz bill proposes a disincentive for a pipeline to place increased rates in effect subject to refund. Repayment of any sums collected by pipeline companies in excess of rates finally approved by the FERC would be made at an interest rate at least three percentage points above the prevailing prime interest rate. Thus, pipelines would pay higher interest to their customers than under present law.[73]

The Fair Marketing Bill, "Natural Gas Fair Marketing Act of 1983" (S. 823: Sen. Roger W. Jepsen, R-Iowa; H.R. 2164: Rep. Thomas J. Tauke, R-Iowa). This bill provides for the decontrol of gas as scheduled under the NGPA. A 50 percent limit is placed on take-or-pay clauses. Failure to reduce purchases of such gas creates a presumption of an "abusive" purchasing practice. "Abusive" would include imprudence or a contract not responsive to market conditions. Unless otherwise determined by the FERC, indefinite price escalator clauses would be declared void.

The Fair Marketing bill would permit the pipeline to "escape the contract" with a producer or negotiate a lower price if the gas is not marketable at the present price under the contract, unless otherwise determined by the FERC. It would permit a pipeline to exercise, without any obligation to pay, a volume adjustment so that it may decline to accept any portion of gas it would otherwise be obligated to accept if it is unable to market such gas. This reduction may be made to a level of 50 percent of contracted volumes. Under the Fair Marketing bill, failure to exercise this volume adjustment right would create a rebuttable presumption of "abuse," as that term would be redefined for purposes of purchased-gas adjustment proceedings.[74] A pipeline could not reduce volume under a particular contract, however, unless it first reduced volume of higher-priced gas to the extent permitted by the bill, and reduced gas of equal or greater price from affiliates. The FERC could determine that reductions should not be made in volume because of field drainage or similar requirements. The pipeline could be required to provide transportation for any volumes of gas that it failed to take.

"Abuse" under the NGPA would be redefined to include: (1) misrepresentation, (2) imprudence by the pipeline, (3) failure by the pipeline to bargain at arm's length, and (4) operating under or entering into a purchase contract that prevents the pipeline from responding to changes in demand or "other market forces."

The Fair Marketing bill would require the FERC to encourage: (1) the restructuring of the purchasing relationship between pipelines and local distribution companies in order to place greater reliance on "competitive market forces," and (2) local distribution companies to seek the least expensive natural gas supplies. Like the administration's bill and certain others, the Fair Marketing bill would eliminate the incremental pricing requirements established by Title II of the NGPA.

The Collins Bill, "Natural Gas Consumer Relief and Market Correction Act" (H.R. 2012: Rep. Cardiss Collins, D-Ill.). The Collins bill is consumer oriented. It would repeal the decontrol provisions in the NGPA and recompute ceiling prices for wellhead sales for certain categories of gas to eliminate price increases in excess of the inflation adjustment since 1977. Take-or-pay clauses would be declared against public policy and void. Pipelines may only pass through purchased gas costs on an affirmative finding by the FERC. Natural gas imports would not be authorized unless the price reflects current natural gas prices.

The Collins bill would place a price ceiling at NGPA section 102 (relating to new natural gas) levels in effect for high-cost gas. This bill would permit the FERC to establish a rate for high-cost tight-formation gas at a level necessary to provide reasonable incentives for its production, but not above the section 102 price.[75] The Collins bill would eliminate the commission's authority to prescribe increased ceiling prices for old natural gas and certain other categories of gas.

The Collins bill would void take-or-pay clauses in all existing and future contracts. In the case of an existing contract the pipeline purchaser may elect to: (1) exercise make-up or similar rights, or (2) receive a refund for amounts paid for gas not yet received. The Collins bill also contains provisions similar to those in the Fair Marketing bill, whereby a purchaser[76] may reduce volumes it would receive to a level of 50 percent of contracted-for deliveries if the purchaser is unable to market that gas. In exercising this option, the purchaser would be required to reduce first the highest-cost gas and then its own production. In addition, the Collins bill provides that any purchaser exercising its volume adjustment rights under the bill shall not, during that period, accept delivery of gas at an equal or higher level under a new or renegotiated contract. Unlike the Fair Marketing bill, the Collins bill would not make a purchaser's failure to exercise its volume adjustment rights an abuse under section 601(c) of the NGPA. Instead, this bill requires the FERC to take into account "the use or failure to use" such rights by a pipeline as an element in any general rate proceeding for that pipeline under sections 4 or 5 of the NGA.

In such a general rate proceeding, the FERC could require the pipeline's shareholders to bear the entire cost of a particular transaction which is not found to be cost effective. This bill also would prohibit the automatic pass-through of purchased gas costs by a pipeline until the commission has investigated the proposed pass-through and, after opportunity for hearing, determined that the acquisition costs that the pipeline is seeking to recover are "just and reasonable and . . . in the public interest."

"The Gas Reform Act of 1983" (S. 1408: Sen. John Melcher, D-Mont.). This bill would establish an interim gas cap based on no. 6 residual fuel oil prices based on the three most recent months less cost of transportation of natural gas to electric power plants. Gas that is being sold at or below this cap on the date of enactment would be "released," when take-or-pay requirements have been reduced to 70 percent of deliverability, proportionately each month over the forty-eight months until complete decontrol. Released gas would be priced at the applicable maximum under the NGPA, or if none, at the lower of the contract price or the gas cap price.

Imported gas, including LNG, and other gas that is sold for more than the gas cap price on the date of enactment would be phased down over a period of thirty months employing proportional assignments of gas to be bound by the gas cap price. Gas not bound by the gas cap price in any month before complete deregulation in the thirty-first month would be sold at the maximum NGPA price or the contract price, whichever is applicable. The price for gas sold under contracts effective after the date of enactment or that are replacements of lapsed, terminated, or expired contracts would be the price agreed on by the parties, except that the gas cap price would set an upper price limit until the thirty-first month. Any party to a contract could terminate it beginning the forty-ninth month and thirty-first month, depending on the date of deregulation applicable.

In the event of reduction in takes under natural gas contracts, the bill would require that, to the extent permitted by state regulatory authorities, all domestic supply contracts should be reduced proportionately. This requirement would not apply to the production of casinghead gas or to reduce production required to prevent drainage or protect correlative rights, or to the extent production is necessary to prevent damage to the well or reservoir so that remaining recoverable reserves will not be reduced. The bill includes provisions for determining the royalties payable to mineral right holders whose royalty is based on market value.

The authority of the president or the Congress to reimpose price controls would be repealed. Provisions of the Fuel Use Act restricting the use of gas by industrial users would be repealed as would the incremental pricing provisions of the NGPA. Title III on removal of impediments to the interstate movement of gas dealing with such matters as the assignment of surplus gas and intrastate pipeline access to interstate supplies previously unavailable to them is identical to the comparable provisions of the administration's bill. The contract carriage provisions of this bill are substantially identical to those of the Senate committee bill.

Bills Dealing with Imported Natural Gas and LNG. The administration's bill does not address the question of imported natural gas or LNG. Bills other than those previously described that deal with imports include the following. H.R. 1441—entitled the "Natural Gas Import Policy Act of 1983," introduced by Representative Tom Corcoran (R-Ill.) on February 13, 1983 would require, effective nine months after the date of its enactment, the suspension of any order authorizing importation of natural gas that was issued pursuant to section 3 of the NGA prior to the effective date of the bill, unless certain requirements have been satisfied.

The Import Policy bill would require that the price of the import contract be renegotiated by the parties to a level responsive to the current market for natural gas and to any periodic changes in the demand for gas during the contract's term. This renegotiation would take place notwithstanding any policy regarding uniform border prices, such as those approved by the Canadian National Energy Board for gas exported to the United States. The Import Policy bill would permit the recovery of certain terminal or attendant facility costs in connection with the importation of Algerian LNG which may be suspended under the other provisions of the bill.

The Import Policy bill would permit the FERC to allow the recovery of costs "prudently incurred" in the construction of the facility. These could be recovered to the same extent as if the facility were used or useful for the importation and transportation of natural gas, without regard to the suspension of the section 3 importation authority. The FERC may not, however, permit any rate of return[77] on such costs to the importing company.

Another legislative proposal concerning imported gas, and imported LNG, is S. 370, the "Imported Liquefied Natural Gas Policy Act of 1983," introduced by Senators Charles Percy (R-Ill.) and Alan J. Dixon (D-Ill.). This proposal would limit the just and reasonable rate for the acquisition of regasified imported LNG to a point not exceeding the average price for no. 6 fuel oil as determined by the secretary of energy during a recent ninety-day period. The Imported LNG bill would allow the FERC and the secretary of energy, upon application filed by the importing company under section 3 of the NGA, to establish a just and reasonable rate for regasified imported LNG in excess of the price based upon average prices of no. 6 fuel oil if: (1) alternative domestic supplies of gas are not available to the market to which the imported LNG is being delivered; (2) the source of supply of the LNG is "reasonably secure from political and technical interruption"; and (3) the LNG supply agreement provides for a reduction in quantity of the LNG or a reduction of its price or both if alternative domestic supplies at reduced prices become available.

4

Setting for the Debate over Wellhead Price Decontrol

The wellhead price of natural gas was not controlled until the U.S. Supreme Court concluded in 1954 that prices of gas dedicated to interstate markets should be controlled under the NGA. The natural gas market is characterized by significant distinctions from the markets for most goods and commodities. Natural gas pipeline companies that move gas from the wellhead to distant markets are not simple transportation companies as is true of railroads and trucking companies. The natural gas pipelines actually purchase most of the gas that they transport in bulk. They in turn sell purchased natural gas to distribution companies and to industrial users.

Pipeline purchase of natural gas is usually accomplished through long-term supply contracts negotiated with producers of gas. Long-term supply contracts are employed to ensure volume deliveries of a steady stream of natural gas. This enables the pipeline companies to guarantee deliveries to gas distributing companies while spreading fixed pipeline costs over a greater volume of gas. In addition, long-term contracts of purchase and sale of natural gas by the pipelines enable them to borrow the funds needed to build and extend their expensive facilities.[78] Pipelines, thus, are major participants in planning for the continuity of the natural gas supply.

In earlier years long-term gas supply contracts often called for producer delivery of quantities of natural gas at relatively low prices. These prices were enforceable against producers even though higher sale prices might be permitted under the federal regulatory regime.

Price restrictions under the regulatory philosophy called for by the U.S. Supreme Court dampened producers' enthusiasm for exploration and production of natural gas for the interstate market, but uncontrolled intrastate gas markets had ample supplies of gas when scarcity of natural gas in the interstate market forced service curtailments in the 1970s.

The natural gas shortages in the interstate market in the 1970s forced pipeline companies to scramble for supplies to meet commitments to gas distributing companies and to industrial users. Producers were then in a better bargaining position and insisted on various contract clauses permitting the escalation of contract prices when regulatory ceilings would permit. Take-or-pay provisions guarantee producers payment on schedule for a large percentage of the gas under contract even though there is no make-up until later or the gas is never taken.

The interstate natural gas shortages of the 1970s and the pricing of oil by the Organization of Petroleum Exporting Countries forced Congress to engage in a major review of energy issues. In 1978 Congress, through the NGPA, ended the dual natural gas market by imposing wellhead price regulation on natural gas for intrastate markets. Provision was made for gradual deregulation of some natural gas prices, complete deregulation of some new gas, and for regulatory approval of increased wellhead gas prices for certain categories of gas. This was done in the attempt to ameliorate the shortages that forced curtailment of natural gas services. Old gas, however, remained controlled as to price and was to be forever controlled. In sum, the regulatory purpose in simple terms shifted primarily from protecting consumers from higher prices to protecting consumers from natural gas shortages. The determination not to be caught short of gas in a future gas shortage period led many pipelines to sign long-term contracts with producers that, in the light of more recent events, seem improvident.

The deregulation of prices authorized by the NGPA encouraged producers to explore for and produce significant volumes of natural gas. Thus, supplies became plentiful again and, predictably, some producers concentrated greater resources on the discovery and production of relatively high-cost natural gas. Some deep gas has sold for as much as $9–10 per thousand cubic feet (Mcf). The increased costs of this gas are passed through to consumers in price averaging, with little regulatory surveillance in purchased-gas adjustment proceedings.

Similarly, the roll in of high-cost imported natural gas and LNG has increased consumers' energy costs. The NGPA provided for "incremental pricing" which sought to shift the increased cost of new supplies of natural gas primarily to industrial users, thus sparing household users. Industrial users and utilities through their planning to minimize fuel costs and, in part, through the stimulus of the Fuel Use Act have switched, in many cases, to the use of alternative fuels. As industrial users have switched to other fuels, gas distributing companies have had necessarily to pass on more of their increased gas costs to household users.

Household consumers accustomed to gas prices below true market value have felt the sting of higher fuel prices and have made their feelings known to members of Congress, thus adding another dimension to the debate over deregulation. The backout of industrial users has helped produce a surplus of natural gas, and pipeline companies have, in some cases, refused to honor their take-or-pay commitments with producers. In former years, householders could use as much gas as they wanted at relatively cheap controlled prices. Increased gas prices in more recent years have encouraged household consumers to conserve gas, and this and the business slowdown have left the pipelines with a surplus of gas.

When the NGPA was adopted, interstate pipelines were at a disadvantage in competing with intrastate pipelines. The NGPA then subjected the intrastate pipelines to regulation, and these pipelines with a smaller cushion of old gas are now viewed by many industry observers as being at a disadvantage in competing with interstate pipelines.[79]

Advocates of wellhead price deregulation believe that the cushion of surplus natural gas at this time makes this a good time to undertake complete deregulation.[80] This is particularly so they claim, because world oil prices are weak and oil prices may set the upper price limit for the price of a Btu equivalent volume of natural gas, since these fuels are substitutable. But opponents of deregulation fear that gas sales contract provisions and what some regard as less than fully competitive markets will produce a fly-up of gas prices or a price spike that may seriously disadvantage consumers.

What are the alternatives for deregulation or a reregulation of wellhead prices? Without repeating the detail involved in the pending legislation, one can summarize the principal thrust of the available choices as follows:

• Congress can do nothing, in which case approximately 60 percent of the natural gas market will be deregulated in 1985 under the terms of the NGPA; old gas will remain permanently regulated.

• Congress can completely deregulate wellhead prices, either immediately or over time, and repeal the Fuel Use Act (which prevents many large industrial users from burning gas) and incremental pricing.

• Congress can deregulate wellhead prices and concurrently modify or abrogate various contractual clauses in gas supply contracts in the attempt to smooth the transition to an unregulated market.

• Congress can impose additional gas price regulation at the wellhead.[81] (Congress could, of course, try to deal with contract clauses and the purchased-gas adjustment even if it takes no other action.)

Each of these choices will be considered in the chapter that follows.[82]

5

Deregulation of Wellhead Prices

Conditions of Competition

Some opponents of deregulation contend that natural gas pipelines enjoy returns to scale on their operations that, but for regulation, could lead to private monopoly. Opponents note that in earlier years representatives of the Federal Trade Commission testified that the natural-gas-producing industry was monopolistic and should not be permitted to exploit consumers by charging monopoly prices.[83] Similarly, in the *Permian Basin Area Rate Case* the FPC claimed that the price of gas controlled by producers was so restricted in relation to demand that the producers had power to negotiate for prices that would be injurious to the public. Some large firms, it is estimated, most of which are integrated oil companies, produce half of the natural gas.[84]

A Congressional Budget Office (CBO) background study suggests that we need to determine whether vertically integrated companies both produce and transport gas for high prices exacted by their production subsidiaries, thus permitting monopoly profits to be transferred from regulated pipeline operations to the producing companies.[85] Pipeline companies are suspect, it is claimed, because some gas producers can sell to only one pipeline and homeowners and other consumers can usually buy only from a source that is served by one natural gas pipeline company. Thus the pipeline may be a monopsony buyer in the gas field and a monopsony seller at the other end.[86]

Howard D. Marlowe, associate director of the American Federation of Labor–Congress of Industrial Organizations, contends that the monopoly power of gas producers makes it impossible for private negotiations to result in fair prices. Thus, in his view, the federal government should protect the public welfare by continuing to regulate the natural gas industry as a utility.[87] An FTC staff report claims that the top eight producing firms account for 42 percent of production; the top twenty, 64.9 percent; and the top thirty, 70.4 percent.[88]

Advocates of deregulation contend that there are well over 5,000 producing firms, including two dozen integrated oil firms, selling gas to pipelines. There are approximately 500 large independent producers and about 5,400 small independent producers. There are only a handful of producing affiliates of pipeline companies.[89] Any conspiracy to withhold gas from the market to exact a higher price requires a manageable number of conspirators. There are simply too many producers to justify speculation that there could be a conspiracy.[90] The concentration ratio in the gas production industry is simply too low to be a concern, according to advocates of deregulation. Indeed, the concentration ratio is considerably lower than for most of the manufacturing industries in the United States. The U.S. Court of Appeals for the Fifth Circuit, moreover, held as early as 1970 that the gas production market is structurally competitive.[91]

The claim of excessive bargaining power on the part of pipelines is also rejected by proponents of deregulation. Although the presence of only one pipeline in many gas fields was a concern many years ago, an increasing number of pipelines compete for gas reserves and there are thirty-five gas transmission companies.[92] Even assuming for the sake of argument that the gas industry were noncompetitive and vertically integrated, there would still be interfuel competition, particularly from oil.[93] The fifteen hundred gas distributing companies would not be affected by deregulation. They would continue to be regulated by state authorities.

Some scholars contend that whether the domestic gas industry is competitive or not is no longer relevant. All uncontrolled energy prices are under the dominant influence of the OPEC cartel.[94] Congress, in phasing out price controls over much gas through the NGPA, sought to phase up gas prices to the level of heating oil, a fuel whose price is largely set by the OPEC cartel. Gas industry companies could not, it is claimed, charge significantly more than heating oil prices in any event. The pipeline companies do not have unchecked ability to pass on increased costs, the CBO maintains, for the higher rates to consumers associated with these increased costs soon lead to reductions in natural gas demand by industrial users.[95]

Residential users, according to opponents of deregulation, generally cannot afford to switch to other fuels, and these consumers will be caught by higher prices.

Price and Supply Responses

A major concern of energy policy is whether there will be adequate supplies of fuels in future years and how effectively increases in prices under various policy choices will stimulate supply responses. Some scholars are concerned that economically recoverable supplies of natural gas may not be large enough and that, in any event, consumer use of natural gas may be relatively price inelastic. The answers to these concerns are critical to the evaluation of the merits of proposals for the deregulation of natural gas prices.

Projected Supplies of U.S. Natural Gas. Estimates of potential gas supplies available domestically vary greatly depending on the economic and technical feasibility assumptions that are made. If there are significant future gas supplies that can be brought to market, the policy debate can focus on the energy strategy that will best tap these supplies and use them efficiently. The quantities of gas now available obviously affect the price of gas. This section looks at the estimates of natural gas supplies available domestically as distinguished from supplies in the world at large or in non-Communist nations generally.

Natural gas consumption in the United States in recent years has ranged up to 20 trillion cubic feet (Tcf) per year.[96] Some officials expect demand for natural gas to be relatively flat during the rest of this decade. The senior energy analyst at the Department of Commerce expects our usage to be 17.9 Tcf in the year 2000.[97] Projections from six sources, stated in quadrillion Btu (or "quads"—somewhat less than a Tcf), estimate total domestic production in the year 1990 will range from 16.1 to 22.5 quads.[98] Projections in recent years of the potential domestic natural gas supply range from 496 Tcf to 1,066 Tcf.[99]

Estimates of the number of years that we can continue to use natural gas at the current rate of consumption range from fifteen years to over a hundred years.[100] One study concludes that "it is not out of the question that oil and gas resources, especially gas, will be discovered on a large enough scale within the coming decade to keep fuel prices from rising at all."[101] Most students of energy issues believe, however, that energy supplies will be more constrained and future supplies will be more expensive.

Elasticity of Supply. Beyond the question of whether there is a future supply of natural gas at any price is the issue of how the supply of natural gas available for consumption would respond to increased prices. Advocates of continued regulation of wellhead prices contend that the supply of old gas is price inelastic. This was the argument of Alfred Kahn in the *Permian Basin Area Rate Case* hearings. Thus, once a quantity of natural gas is committed to long-term contract, that quality of gas and the prices to be charged for it are fixed. The costs of discovering and developing the gas have already been incurred, and costs of continued production will not change greatly. The quantity of gas subject to the contract will not change with later changes in the market price of gas. If the regulation of these prices is relaxed, the producer, it is claimed, will gain "windfall profits." Continued regulation of prices will minimize windfall profits and not interfere with the flow of the supply of gas committed to that contract.[102]

Similarly, the General Accounting Office concluded in 1976 that few additional gas reserves would be discovered after an increase in the price of gas.[103] The Department of Energy argued in 1977 and 1978 that gas supplies were relatively inelastic with respect to price.[104] Consumer representatives then argued that ending wellhead price regulation would mean a transfer of wealth from consumers to producers without any compensating economic or social benefits. In their view, econometric models had not reliably demonstrated that the supply of natural gas was elastic.[105]

Advocates of deregulation contend on the other hand that the results of studies with models, particularly the more recent studies, show that gas supply would be responsive to price increases.[106] Advocates contend, moreover, this demonstrates that the natural gas supply is price elastic. In fact increased gas reserves were discovered when the NGPA deregulated some categories of gas and provided for gradual deregulation of some other categories. As a corollary, the gas shortages of the 1970s were produced by a policy that wrongfully considered natural gas supplies to be price inelastic. Producers would not explore for and develop new gas reserves when prices were strictly regulated, but lifting some of these price strictures produced a great increase in gas exploration. This added substantially to future supply and discredited the conservative supply estimates advanced by the Carter administration. Advocates of deregulation argue that low prices will produce a gas shortage but higher prices will relieve the shortage.[107]

Richard J. Pierce, Jr., notes that old gas has a lower price elasticity of supply than new gas, for example, but this does not mean that old gas is price inelastic. He argues that every study of price elasticity for various forms of energy conducted in the past few years has seriously underestimated the responsiveness of supply to price increases. All wells of old gas, moreover, do not have the same marginal costs of production. Thus, prices set by regulation may exceed the cost of production in some cases but understate the cost of production in other cases. Rigid price limits for various categories of gas discourage efforts at secondary recovery. As a result, much old gas will simply never be brought to market as producers concentrate on producing unregulated gas or gas subject to less restrictive price regulation.[108] A major oil company argues that freeing old gas from price controls would yield the equivalent of three years' supply of natural gas at current consumption rates.[109] Other estimates of the additional gas that would be available with the decontrol of old gas range from 5 to 52 Tcf.[110]

Effect of Contract Clauses on Prices. Sales of natural gas by producers are generally consummated pursuant to long-term contracts. A contract to deliver a specified volume of natural gas for a stipulated price without escalation would, of course, be enforced against the producer even though the stipulated price is less than the government's regulatory regime would permit. Producers, to avoid being locked into prices below the permissible level under regulation or in anticipation of a deregulated market, have included various contract provisions in their gas supply contracts. Pipelines have insisted less frequently on inclusion of market-out provisions that will permit them to decline acceptance of some or all of the gas contracted for if higher prices provided for in various escalation provisions prevent them from delivering to customers the quantities initially agreed upon with the producer.

The principal types of contract clauses that producers have insisted upon include the following. Fixed-price escalation clauses, such as an automatic increase of a fixed amount in price at stated intervals, are included in perhaps 8 percent of the supply contracts. The parties to these supply contracts recognize that, when a gas well is initially placed in production, subterranean pressure is high and gas will flow easily. In time, however, the producer may have to pressurize the subterranean area and rework and clean the well. Escalation of prices anticipates these costs though, of course, these costs are not always incurred. Pipelines are given the benefit of low initial prices when they have high marketing outlays. Higher prices are imposed later under the fixed-price escalation contract when the pipeline has completed its marketing work and has recovered a significant part of its capital through depreciation.

Producers claim that indefinite price escalator clauses (see glossary) are employed to offset anticipated inflation and the possibility of increased taxes and other producer expenses. Critics contend the operation of these clauses is unrelated to these contingencies. Perhaps 56 percent of the pre-1977 contracts for sales to interstate pipelines had most-favored-nation clauses, and most of these contracts executed since April 1977 have such clauses. These clauses enable producers to capture increased income if increased demand for gas produces higher gas prices.

Perhaps 15 percent of the supply contracts have price redetermination clauses that spell out the conditions under which either party to the contract may reopen price negotiations. These provisions generally specify when negotiations may be reopened and often state the method of arbitrating price. One survey estimated that 51 percent of the price redetermination clauses actually specified price in the event of redetermination.

Many supply contracts have take-or-pay provisions under which the pipeline agrees to pay for a specified percentage, averaging 80 percent of deliverability (total contract volume), whether the pipeline actually takes that volume of gas or not. Many take-or-pay provisions have make-up clauses allowing pipeline purchasers a period of time, such as five years, in which to take the required percentage that is initially rejected. These clauses benefit pipelines, according to their defenders, by enabling them to avoid taking some percentage of the total gas they have agreed to take when changed market conditions make this advantageous.

Take-or-pay clauses, according to some critics, permit producers to exceed, in effect, regulated prices, since they are paid on schedule for the required percentage, even though the minimum balance is not taken on schedule, and they can earn interest on these advance payments. These critics argue that pipelines signed contracts with these clauses because they wanted gas and felt they could pass the costs on to consumers. The percentages pipelines were required to take was so high the small percentage they could avoid taking did not help consumers. The provisions reduced the pipelines' flexibility in dealing with changed market conditions.

Perhaps 26 percent of the supply contracts have area rate clauses that tie wellhead prices to the highest price permitted by regulatory authorities. This ensures that regardless of the initial price agreed on, producers will receive such escalations in price as are permitted by the regulatory regime. Many contracts, of course, have more than one pricing clause that permits escalation of prices.[111]

Clauses such as area rate clauses anticipate continued regulation of wellhead prices. Other contract provisions clearly anticipate pricing in the event of deregulation. Congressional intent as to the treatment of these contract clauses in the NGPA is unclear. The FERC, after its initial order refusing to allow producers to employ escalation provisions in existing contracts to collect maximum lawful prices under the NGPA, reversed this prohibition and stated that parties to pre-NGPA contracts could amend them if necessary to provide authority to pay and receive maximum lawful prices under the NGPA.[112] Pipelines agreeing to these amendments either believed this would enhance their access to future supplies of gas or that the cost could readily be passed on to the consumers.

The impact of contract clauses on natural gas prices in the event of total deregulation is a concern to some policy makers. It is almost of equal concern if the current schedule of deregulation proceeds under the NGPA so that a majority of available gas is completely deregulated in January 1985.

Many observers fear that various escalation provisions, either singly or in combination with other provisions, will drive gas prices above long-term and partial decontrol market-clearing prices. Some clauses, for example, call for prices 110 percent of no. 2 fuel oil.[113] Even a small number of oil-tied clauses could, through application of most-favored-nation clauses, force the average price above market-clearing levels. Take-or-pay provisions may have such high minimum take-or-pay requirements that pipelines would be unable to reduce average prices quickly.[114] Thus some companies might go bankrupt, and high gas prices would force fuel switching and overreliance on imported oil.[115] The contracts issue is as important, according to some officials, as the issue of deregulation of old oil prices.[116]

Other observers believe the contracts problem may be overstated. Some observers believe that producers and pipelines, when faced with the hard realities of the situation, will negotiate their differences and arrive at prices for deregulated gas that will not exceed market-clearing prices. Thus the market value of gas rather than contract provisions may in actuality set gas prices.[117] There are circumstances that may militate against a serious crisis in gas pricing. There are, for example, some though not a large percentage of contracts tied to oil prices that have market-out provisions. Many most-favored-nation clauses refer to an average of three contracts with similar clauses. Some oil parity contracts have transportation netback provisions that have the effect of reducing gas wellhead prices.[118] Area rate clauses, unless backed up by other clauses more favorable to producers, may yield prices below market-clearing prices after deregulation.[119] Thus, even if prices under some contracts increase, pipelines will be able to pay less than market-clearing prices on enough gas delivered under contracts with more flexible provisions to keep average prices from being a serious problem.

Various legislative remedies have been proposed to ameliorate the contracts problem. They range from nullification or modification of certain contract provisions to lowering the minimum-take requirements under take-or-pay provisions, to placing a cap on all regulated gas on the date of legislation (such as a percentage of oil prices or an average of gas prices newly contracted for delivery in the area), to allowing liberal market-out opportunities, to doing nothing.

Advocates of government action to effect changes in contract clauses and permit more liberal market-out privileges contend that without government action contract clauses will cause gas prices to fly up unreasonably. If contract clauses are only changed as old contracts expire and new contracts for the delivery of gas are written, this will not have sufficient impact to prevent a price spike. Charles J. Waidelich, chairman of the board and chief executive officer of Cities Service Company, notes that many gas contracts were signed years ago to comply with regulatory requirements that no longer apply. He regards the market-out and renegotiation features of the administration's bill as the most ingenuous features of the bill. Legislative action is required, in his opinion, because otherwise mineral owners might, under the terms of their gas leases, cancel the leases to the detriment of the producers.[120]

Charles Cleveland, director of the Coalition for Fair Utility Rates, argues that abolishing take-or-pay clauses and escalator clauses, and prohibiting their use in the future, is needed to lower fuel costs. Pipelines should be required to obtain the lowest-priced natural gas.[121] Howard Marlowe, associate director of the AFL-CIO, argues that limiting take-or-pay clauses would let pipelines take advantage of lower-priced gas.[122] If contract clauses are not modified, there will be uncertainty in the market because future prices cannot be determined until the time the price changes under indefinite price escalation clauses become applicable. If deregulation produces a price fly-up and gas prices exceed market-clearing prices, some consumers, particularly industrial consumers, will switch to other fuels. This will stimulate increased oil imports to the detriment of our balance-of-payments account. Some gas producers and pipelines would go bankrupt.

Opponents of government intervention contend that even though Congress may possess the power through regulatory legislation to modify contractual arrangements between producers and pipelines, it should avoid such action. Opponents contend, first, that many producers and the pipelines are already taking steps to renegotiate contracts. Douglas W. Williams, senior vice president of Lone Star Gas Company, points out that his firm has obtained voluntary price concessions from thirteen high-priced suppliers. Only one supplier has refused to make unilateral reductions in price. The concessions are made, according to Williams, because producers want to ensure the continued marketability of their gas supplies. They are concerned about the long-term viability of the natural gas industry.[123] Whitney Marvin, president of Whitmare Exploration Company, points out that Shell, Arco, and Amoco have agreed to limit or reduce take-or-pay requirements with his firm.[124]

Representative Daniel R. Coats (R-Ind.) contends that a total ban on most-favored-nation clauses would lead to short-term supply contracts with greater price volatility.[125]

Arlon R. Tussing claims that universal market out would be monstrously inequitable, and if the contract clauses are too unreasonable, they are unlikely to survive in any event. He argues that the bankruptcy of a single firm in the industry would do more to solve the contracts problem than any action that Congress could take.[126]

Opponents of government intervention fear that wholesale modification or abrogation of contracts might remove all incentives for producers to find and develop new natural gas resources.[127] Banks who extend credit to producers would reexamine the creditworthiness of producers, many of whom are relatively small independent companies. Since these producers would not, with legislative abrogation or modification of contract clauses, have the security of contract on which the banks agreed to provide credit initially, these producers would have their credit severely restricted and some might go bankrupt.

Some intervention proposals would impose serious burdens on the regulatory authorities and the courts to determine prices for natural gas, providing what some refer to as a lawyers' relief act. Gas caps would reregulate portions of the natural gas deliverable under contracts containing the criticized clauses, thus requiring pipelines to roll in prices of unduly expensive gas with gas that is cheaper because of the gas cap. This raises consumer prices notwithstanding the cap and prevents the efficiencies of marginal cost pricing. Why, opponents ask, should gas prices be reregulated when the FTC has found the natural gas production industry to be reasonably competitive?[128]

Probable Price and Supply Responses. Many observers feel that the probable effect of deregulation or partial deregulation of wellhead prices for natural gas after nearly thirty years of regulation will be difficult to predict. Some observers believe that the long-term effects of deregulation many be quite different from the short-term effects. Supply responses may be considerably lagged, according to some scholars, because natural gas exploration and development are capital intensive, new exploration may not be immediately successful, and initial production may increase reserves but not market supplies.[129] This section cites arguments as to the price and supply effects of different approaches to the deregulation or reregulation of natural gas at the wellhead.

Responses with complete deregulation. Total deregulation, it is claimed, would reduce producer uncertainty over future regulatory policy, thus adding to the incentive to invest in exploration for and development of natural gas.[130] There would be increased gas supplies, but this gas might not be available for consumption, however, for some time.[131] Gas prices, according to many observers, would be more uniform from region to region than would be true without complete deregulation.[132] Some scholars contend that the price of gas for industrial users and electric utilities would be near the energy-equivalent price of the grade of fuel oil these industries and utilities could use as an alternative to gas. Suppliers of gas probably would not increase prices to exact parity with that fuel, however, since this might stimulate switching away from natural gas as a fuel.[133]

Petroleum economist Frank W. Lewis testified that consumers are clearly reducing gas consumption as gas prices increase.[134] Christine Hansen, commissioner of the Iowa Commerce Commission, stated that a survey in Iowa showed a 3.75 percent to 6.6 percent reduction in use for every 10 percent increase in price.[135] Repeal of the Fuel Use Act might, of course, stimulate some industrial users and utilities to switch to using natural gas because of its special qualities.

Some observers believe that gas is so abundant domestically and internationally that there may not really be a fly-up or price spike with decontrol.[136] New gas, some claim, is selling below expected price levels. In many places heating oil is displacing natural gas as a fuel because of price advantage.[137] Over 60 percent of gas today is consumed by industrial users and utilities that generally have energy alternatives. Gas prices must be kept competitive with oil prices if these customers are to be retained.[138] Some observers believe that total deregulation, with prices being set by supply and demand rather than by regulation, may in some markets produce lower prices for consumers than would be true under regulation.[139]

William K. Tell, Jr., senior vice president of Texaco, Inc., says that although there were fears that gasoline prices would fly up when crude oil prices were decontrolled (gas being a product of crude oil), in fact gasoline prices began to decline with decontrol and they are now lower, allowing for inflation, than they were when decontrolled in January 1981. Tell argues that deregulation makes competition more intensive at all levels of the industry, and high-cost operations must be eliminated.[140] Tell argues that efficiency benefits from decontrol include removing pipeline incentives to purchase high-cost supply increments, eliminating regulation that distorts price signals in the marketplace, and encouraging the production of the lowest-cost gas first.

As to deregulation of forever-regulated old gas, which is currently priced well below new-gas supplies, industrial users contend that we must eliminate our current multitiered price structure and allow marginal rather than roll-in pricing of energy supplies. This would allow ultimate purchasers, through their actions in the market, to send accurate signals to producers. Currently the regulatory regime and roll-in pricing hide true costs from most consumers. Supply and demand will set prices in the marketplace more efficiently, it is claimed, than regulation. Congress, according to Hubert Gentry, Jr., senior vice president of Entex, Inc., cannot prevent a windfall by retaining price controls on forever-controlled gas. Rather, the windfall is simply transferred to the producers of unusually high-priced new gas.[141] Robert Boswell, president of Adams Energy Corporation, insists that old gas has greater value now than when it was first marketed and current pricing should recognize this.[142]

Total and immediate decontrol is advocated by Kenneth L. Lay, president and chief operating officer of Transco Energy Company, as the cost-efficient procedure, since the existing system of controls and incentives is too complex and confounded to be dismantled piecemeal.[143] Lay contends, moreover, that with the present surplus of gas and lower oil prices, there is less likelihood that decontrol will cause abrupt price increases. Paul W. MacAvoy contends that under the administration's plan old gas will continue to be regulated for a longer period of time and that the gas cap for new gas would be more liberal than the deregulated price level. He believes that competitive factors would bring prices down more rapidly under complete decontrol.[144]

Decontrol would shift the search for and development of natural gas resources from extremely expensive sources to the development of the most easily recoverable gas, thus keeping prices down. If decontrol produced higher gas prices, this would stimulate greater exploration and production. In addition, industrial and household consumers faced with higher prices would take steps to conserve natural gas by purchasing more efficient equipment, installing insulation, and so forth, thus promoting greater efficiency in energy use.[145]

Some opponents of complete price decontrol for natural gas, or decontrol without government action to defuse contract clauses, contend that decontrol may indeed produce a price fly-up or spike. One study that assumed complete decontrol January 1, 1983, estimated that there could be a rise of 20 percent in the average price of gas for residential consumers and 30 percent increase in the average price paid by large industrial users. If oil prices weakened, there would be less risk of a price fly-up.[146] Contract clauses written in anticipation of deregulation would be triggered by the elimination of price controls.[147]

The CBO notes that, historically, short-run energy demand is inelastic in nature.[148] Thus, opponents of deregulation contend that a rapid increase in gas prices would greatly increase consumer outlays for gas. Residential consumers, and industries that require natural gas for glassmaking and so forth, would be caught with these higher prices. If home building activity revives and the business recovery proceeds, there will be added demand for natural gas and this will put additional upward pressure on gas prices.[149]

Opponents of complete decontrol contend that deregulation of old gas will not stimulate the production of additional gas. Price increases cannot create additional gas in the ground; that takes nature millions of years.[150] Wells producing old gas for which initial costs have been recovered are, according to Robert A. Hefner III, president and managing partner of GHK Companies, cash cows that produce income with a minimum of effort. Hefner claims the government would have to abrogate tens of thousands of gas supply contracts to free the price of old gas completely. If prices of some new gas are already high, prices of old gas cannot be raised without reducing the prices of this new gas.[151] Hefner claims that reports of the Energy Information Administration, ICF International, and the Department of Energy's Office of Policy, Planning, and Analysis all concluded that deregulating old gas would produce less additional gas supply than would the continuation of the NGPA. As to variations in gas costs for consumers in different localities, Hefner claims that there are even greater regional differences in the cost of electricity.[152]

Response to decontrol with modification of contract clauses. Stewart notes that prices continue to escalate notwithstanding a substantial oversupply of gas. Contract provisions will compound the problem if prices are decontrolled.[153] Gentry contends that we need to find a mechanism that lets gas prices go down as well as up.[154] The administration's proposal, among others, would modify prior contract clauses to reduce the risk of a price spike from the total deregulation of natural gas prices and to assist the market in adjusting to deregulated conditions.

Advocates of legislative modification of contract clauses insist that without such action too few contracting parties will be able to reach new agreements that will ensure lower prices. Prices will soar, with the risk of fuel switching by industry, which will force some gas companies into bankruptcy.

Legislative action to adjust the terms of contract provisions and permit pipelines to market out will reduce the period of turmoil in the market, it is asserted, and set conditions under which prices will be competitive with other fuels. Producers will tap less expensive sources of supply. Deregulation in the long term will encourage the exploration and production needed to serve gas customers for many years in the future. C. M. Butler, chairman of the FERC, argued in 1982 that deregulation in the current market could, if the contracts problem could be eliminated, lead to a reduction in gas prices.[155]

Opponents of the strategy of coupling legislative modification of contract clauses with decontrol contend that the contract provisions in question were voluntarily assumed by the parties to the contracts. The government above all should observe the sanctity of contracts that were freely negotiated. Eliminating certain of these constraints and allowing parties to, in effect, market out may mean that they will negotiate contracts at higher prices. Prices of old gas thus freed from constraints may rise to two and a half times their current levels.[156]

Impairment of take-or-pay clauses would reward pipelines that made imprudent contracts while penalizing those that acted responsibly. Modifying contractual arrangements would destroy the ability of producers to plan and finance new exploration and development for natural gas. Impairing contracts would also damage the pipelines' ability to attract capital,[157] particularly if they must buy gas in the future on short-term contracts or in the spot market.

Response under proposals for reregulation. Several legislative proposals would continue or reimpose controls over natural gas. Marlowe advocates rolling back prices and extending price controls. He contends that the NGPA has allowed gas prices to escalate far more than justified and more than Congress foresaw. Extending controls would prevent a price explosion and give Congress time to assess what additional action should be taken.[158] Advocates of this approach regard the risk of unduly high costs for natural gas as too great to permit experimentation with deregulation, particularly when household demand is viewed as relatively inelastic.

Critics of additional controls contend that price controls put the government in the position of making economic decisions about energy and about who will be favored with artificially low prices. As soon as the government intervenes and sets prices below market-clearing levels, supplies dry up and a nonprice rationing system must be established.[159] Low-priced gas for some consumers means others will have none as Congress must "allocate the shortages."[160] The system employed to regulate natural gas adds transportation and distribution costs to wellhead price to arrive at retail price, an add-on system under which marketability is addressed at the end of the process instead of at the starting point. The inertia of the system does not permit market signals from consumers to be transmitted backward to establish a market-clearing wellhead price or alter other charges to avoid load loss when customers switch to other fuels because of high prices.[161]

Critics contend that the political reaction to signals that controls on natural gas are a failure is to patch up the controls with more controls, but that only creates new problems. Only an unregulated market, it is claimed, can resolve the ever-increasing conflicts and inconsistencies produced by regulation and resolve market ordering problems.[162] The problem with recontrols, according to Clyde H. Greenert of Union Carbide Corporation, is that controls have never worked. Wellhead price controls led to disastrous shortages of natural gas for some consumers.[163] Recontrol would devastate domestic gas exploration and development by removing incentives for exploration and development while increasing our dependence on foreign energy sources.[164] When gas supplies dwindle, the environment will suffer, since alternative fuels pollute the atmosphere. Many of our undiscovered natural gas resources may never be recovered and used.

Although the administration's proposal calls for deregulation in 1986, prices for categories of gas controlled under the NGPA would be subject either to those controls or to a gas cap if contracts are renegotiated. Critics contend this amounts to reregulation.[165] This gives Congress the opportunity to tinker with gas caps continuously so there may never be deregulation. The gas cap will roll back some prices rather drastically, it is claimed. Some independent producers will be forced into bankruptcy because pipeline companies would not have to honor take-or-pay provisions.[166]

If the gas cap is below the market-clearing price, this will have a chilling effect on exploration, and gas shortages will follow. If the gas cap is set as stated and deregulation in fact occurs in 1986, gas prices will indeed fly up at that time.[167]

Defenders of the administration's proposal contend that there will be no fly-up of prices with decontrol in 1986 because gas prices will then be at the market-clearing price.[168] P. Thomas Cantrell, deputy assistant secretary for energy, acknowledges that the administration's gas cap proposal will hold down prices. He claims that econometric studies show the gas cap is not necessary. It is really a standby safeguard—an insurance policy against the unforeseen.[169]

Responses under the NGPA. Pipelines, a number of producers, and some scholars believe that the best course of action for Congress would be not to change the NGPA. Tussing put it that Congress should observe the first canon of the medical profession, "do no harm." This he equates in the regulation of natural gas to doing nothing at this time except, perhaps, repeal the Fuel Use Act.[170]

Hefner argues that the essence of legislation in a democratic system is compromise and that the NGPA is the best compromise achievable within political realities and a good balance between producer and consumer interests.[171] Advocates of this position note that we will decontrol eventually as supplies of forever-regulated old gas are depleted. In addition, consumers may not feel the full effect of NGPA decontrol of most gas prices for some time because state utility regulatory commissions are slow to grant price increases to gas distributors.[172]

Repeated tinkering with the legislative rules in an attempt to chase gas prices and supply conditions back and forth across the business cycle and to fix transitory problems with grand structural reforms, according to Tussing, are likely to catapult the industry from one crisis to another. First, the increase of prices feared with the deregulation of most gas prices in 1985 under the NGPA has already occurred. Second, several pipelines are already exercising market-out, renegotiation, and arbitration clauses in contracts, and some firms that lack this authority in their contracts are proceeding as though they in fact have such authority. Tussing says "so what" if the prices of some deep gas are soaring to \$8 to \$10 MMBtu. Who, after all, is still buying at those prices?[173] In sum, advocates of nonaction believe that the current turbulence in gas markets will best work itself out if Congress does nothing. Gas prices will find an equilibrium heavily influenced by oil prices, and gas supplies will be forthcoming consistent with the market-clearing price, which is the supply response that would obtain under complete deregulation.

Advocates of deregulation, however, contend that the NGPA extended controls to the intrastate gas market, which now has higher average gas prices than the interstate market. The NGPA contributes to increasing price variations from one section of the country to another. It encouraged accelerated development of very high-cost gas while neglecting lower-cost resources.[174] Gas users at the end of a 2,000-mile pipeline sometimes obtain cheaper gas than consumers next door in the intrastate market.

Advocates of deregulation say we need to maximize efficient supply sourcing. The NGPA, however, provides incentives to recover the most expensive, deep, difficult-to-obtain gas, bypassing shallow, more easily recovered gas. Even though conservation and a troubled economy have reduced energy demand, some gas prices under the influence of indefinite escalator clauses keep rising against the trend, with the regulatory process penalizing the very consumers it was designed to protect. If there is no fly-up in prices under the NGPA in January 1985, this will be because of luck and not because of the NGPA. But if oil prices rise between now and January 1, 1985, there will indeed be a fly-up of gas prices. In sum, Tell argues, the NGPA assures us of neither adequate supplies nor reasonable prices.[175]

The NGPA would not deregulate all categories of gas. Pipelines' incentive to buy high-cost gas to meet demand varies according to the amount of their low-cost gas cushion. The NGPA thus serves to boost the cost of uncontrolled gas while low-cost gas is left in the ground because of high take-or-pay requirements in contracts for the more expensive gas. Perhaps 52 Tcf could be added to our old-gas reserves, say advocates of deregulation, but for the perverse incentives of the NGPA and contract clauses.[176]

Distributional Effects of Decontrol

Many economists and others, on the one hand, are concerned with the efficient operation of energy markets and with prices for natural gas that will encourage production and simultaneously discourage inefficient use of gas. Consumers, on the other hand, often see high energy prices as the problem and not the solution.[177] Although efficiency concerns have forced consideration of changes in natural gas regulation, distributional issues have heavily influenced the changes that have been made.[178]

Complete decontrol, of course, will affect both efficiency and the distribution of wealth. Legislation that modifies or abrogates certain contract clauses will redistribute wealth. Repeal of the Fuel Use Act would enable industrial users and utilities to use natural gas in circumstances in which they cannot now do this, and this will affect natural gas markets. Repeal of the incremental pricing provisions of the NGPA would eliminate lower gas prices for residential consumers at the expense of large industrial users. This section assumes these changes are made and considers the distributional changes that would follow.

Effect on Consumers. Deregulation may have somewhat different effects on consumers in different regions of the country, an issue that will be dealt with last in this section. Consumers will be affected differently if average prices remain relatively level or, alternatively, if prices increase as feared by some observers. Industrial and household consumers obviously will be affected somewhat differently.

Effect with level average prices. With relatively level average prices, additional supply of natural gas would probably be encouraged by the greater certainty in the market. Having settled the ground rules, Congress would let producers and pipelines plan their future operations with some certainty. Banks and others who help finance exploration and development would feel more secure. Producers and pipelines, with the repeal of the Fuel Use Act and incremental pricing, could assume that industrial users and utilities, whose uses are now constrained by that legislation, would consume more gas. More utilities could, for example, burn a mixture of coal and natural gas to meet clean air standards.[179]

Intricate and difficult administration and enforcement requirements for incremental pricing would be scrapped, thus lowering compliance costs for industrial users and utilities. This would raise the rates of residential consumers somewhat.[180] Industrial users whose processes require the use of natural gas would feel more secure that, should shortages of gas develop in the future, increased prices at that time would stimulate increased exploration for gas reserves and increase supplies.

Many industrial users rely on intrastate pipelines for their supplies, but these pipelines are currently at a disadvantage in bidding for additional gas supplies because interstate pipelines have the advantage of a larger cushion of old low-cost regulated gas.[181] Changes accompanying deregulation should, some observers contend, enable intrastate pipelines to compete more effectively for the natural gas supplies available in their states.

Even though residential consumers would lose the benefit of lower gas rates, this loss might be partially offset by the lower prices they would pay for products of industries that would have lower gas rates because of the termination of incremental pricing under the NGPA.[182]

Critics of deregulation contend that we should not repeal the Fuel Use Act because we should be using abundant coal that is available at relatively cheap prices rather than use natural gas. Natural gas should be reserved for high-priority users, such as residential consumers, and for industries whose processes require natural gas, such as petrochemical works. Retaining incremental pricing, it is argued, would keep lower rates for residential consumers. Many of these consumers are lower-income families who would especially feel the burden of higher prices. Industrial consumers, however, can factor in higher energy costs under incremental pricing and pass these on in the prices of the products they sell.

Effect with increased gas prices. Some observers believe that, with only about 25 percent of natural gas being consumed in the residential market, the price of gas in the decontrolled market will be primarily determined by competition for sales in the business sector.[183] In that sector, gas competes with fuel oil and with coal, with prices likely to be heavily influenced by the price of fuel oil. If gas prices increase, this will probably stimulate substantial fuel switching by industrial users and utilities to alternative fuels. These consumers, it is claimed, can often switch to alternative fuels in a rather short time if price adjustments make this desirable.

The Energy Information Administration reports that 70 percent of the gas consumed by large industrial boilers in 1979, representing perhaps 16 percent of all gas consumed by industry, could have been dropped in favor of alternative fuels with minor adjustments to boilers.[184] Domestic consumption of natural gas for 1982 was down 8 percent from 1981. Over 80 percent of this decline was in industrial consumption, which declined by 17 percent over this period of time.[185]

Energy Secretary Donald Hodel fears that a loss of industrial consumption would make the remaining consumers, both industrial and residential, vulnerable to increased burner tip prices as fixed costs of gas distribution systems would have to be borne by fewer users.[186]

Opponents of deregulation contend that higher gas costs would be particularly hard on lower-income families.[187] The capital cost for households to shift to alternative fuels is quite high, in excess of $1,000 per home according to one estimate.[188] Lower-income families would thus have few alternatives for escape from higher gas prices.

Looking at short-term direct effects of deregulation and assuming a gas price increase, the American Gas Association concludes that socioeconomic groups with a higher-than-average proportion of low-income families would suffer proportionately more. The gas bill for families headed by a black person would increase from 3.1 percent to 5.5 percent of family income. Families earning less than $5,000 in 1977 income would spend an additional 8.4 percent of their incomes on utility bills, and those with incomes between $5,000 and $10,000 would spend an additional 3.3 percent on utilities.[189]

Higher gas prices, it is claimed, would extract substantial additional sums from consumers of natural gas and transfer that money primarily to producers of natural gas, since pipelines and distributing companies that transport the gas are regulated utilities whose rate of return is limited. The size of this wealth transfer depends, of course, on the magnitude of price increases, the conservation measures taken by consumers, and the amount of fuel switching that takes place.

Dampened demand for natural gas, it is claimed, would retard exploration and development for new gas supplies, offsetting the stimulus of higher prices.[190] Higher gas prices might increase reliance on imported oil. The amount of any increased oil imports would, of course, depend on the relative price of natural gas and imported oil and the extent to which coal could capture part of the boiler fuel market.[191]

Advocates of gas price deregulation contend that for some industries natural gas is irreplaceable. The chemical industry alone uses 10 percent of all gas consumed domestically. Products of natural gas are ingredients in countless products from medicines, to pesticides, to car parts. Fully half of the nation's business is at least partially dependent on petrochemicals.[192] Substantial industrial use of natural gas will continue. Repeal of the Fuel Use Act would enable many additional firms to switch to natural gas if they wish. Thus, the decline in industrial use may be much less than expected in some quarters.

If higher prices result, this, it is claimed, will lead to wholesome conservation measures and less wasteful use of natural gas.[193] Some scholars contend that each 10 percent increase in energy prices will induce conservation of gas equivalent to 2 percent of current consumption in the short run, increasing to 5 percent in the longer term. If some consumers reduce consumption, this will make available additional supplies of gas to meet excess demand in the northern part of the United States. The dollar gains from the prices that these consumers would be willing to pay for gas they might not otherwise be able to obtain in a period of scarce supplies, according to the 1976 study of Paul MacAvoy and Robert S. Pindyck, would exceed the losses inflicted on those who would no longer receive gas at subsidized prices.[194]

As to the claimed adverse effect of increased natural gas costs on lower-income families, scholars acknowledge that families with incomes below the median spend a somewhat higher percentage of their incomes on fuel. A Massachusetts Institute of Technology study group, however, concluded that consumption of energy through purchase of goods that are produced by industry with the use of energy is twice as important as direct consumption of energy. That group concluded that a more refined study might show that energy consumption, direct and indirect, is the same fraction of income for all groups and does not have an adverse effect on the distribution of income.[195]

Other economists draw similar conclusions. Jack Stockfisch of the American Petroleum Institute concludes that "the net effect of gas price deregulation on the distribution of income is ambiguous. No clear case can be made that these effects will be regressive."[196] Ronald Schmidt of the Federal Reserve Board of Dallas analyzes direct and indirect effects of gas price increases and concludes that such increases would not transfer income from the poor to the wealthy, noting, among other things, that industries would shift a greater portion of their production inputs to labor.[197]

Some scholars contend that, in any event, use of energy policy is the wrong way to achieve equity in income distribution and living standards. Distributional goals are better attained through tax and transfer policy.[198] Indeed, the income distribution effects of energy price increases are already constrained, it is claimed, by income maintenance programs and by the progressive income tax structure. Most income maintenance programs automatically increase benefits when price indexes rise, leaving those affected about as well off as before.[199] Income tax burdens for these energy users are, of course, much lighter.

Effect on Pipelines and Producers. Deregulation and modification of contract clauses would, in time, eliminate the cushion of old regulated gas that permits some pipelines to bid up the price of high-cost new gas and thus gain greater gas supplies.[200] Interstate pipelines currently hold contracts for much greater quantities of old controlled gas than do the intrastate pipelines.[201]

Deregulation should, in time, move the market toward marginal cost pricing. Thus gas supply contracts signed initially or renegotiated after enactment of deregulation legislation should adjust the price of old gas substantially upward. Prices for natural gas at the wellhead should be more uniform throughout the United States than under the NGPA, which perpetuates great disparities in gas prices.

Pipelines with a large cushion of old price-regulated gas are more likely to oppose legislative modification of contract clauses, whereas those with a small cushion of such gas may prefer modification so they can market out of high-cost contracts and seek cheaper supplies.[202]

Deregulation of natural gas prices would affect many producers quite substantially. Companies owning old price-controlled gas would stand to gain significantly from the deregulation of prices. A considerable portion of the supply of old controlled gas is owned by major oil companies. Producers of high-priced currently deregulated gas would stand to lose considerably as their supplies of gas are now often priced above the market-clearing price for natural gas. Some aggressive independent firms have done a great deal of the exploration for high-priced gas. They believe that deregulation would in effect transfer income that they might have earned to the firms that own old gas, who would be able to increase prices on that commodity.[203]

An unreleased EIA report concludes that if average gas prices increase from the present average of $2.69 per Mcf to $3.20 with decontrol, the larger oil companies, because of their greater holdings of old gas, would gain $4.2 billion in revenues the first year after decontrol. Even if prices remain stable at $2.69 after decontrol, these firms would gain $1.85 billion in revenues.[204]

Regional Differences with Decontrol. The effect of deregulation on gas prices and consumption in different regions of the United States depends significantly on the price of the fuel with which natural gas must compete in the particular region. Some scholars have ranked regions by the differences between their oil and gas costs to determine how much consumers are willing to pay for natural gas. They conclude that, in general, regions, under decontrol, would shift toward more use of the fuel that was principally used before decontrol.[205]

In the case of New England, economists Lynn E. Browne and Richard F. Syron conclude that deregulation would mean substantial conversion from the use of natural gas to oil but a narrowing of the energy price differential between that region and the rest of the country. On the basis of 1974 prices used in their study, these economists concluded that energy costs in New England would be one and a half times those in the rest of the country rather than 220 percent.[206] New England has less than 4 percent of the nation's residential gas customers as against 5.7 percent of the housing units. Thus, the negative effects of decontrol would be less than elsewhere, and deregulation would improve the competitive position of New England's industry by reducing the differential between energy costs in that region and elsewhere.[207]

An analysis by Robert Leone concluded that gas decontrol would mean a net gain for all states in the northeast region except Pennsylvania. The net gain for the region stems in part from improved competitiveness of industries in the region. Profits would remain stable or tend to increase because there would be less difference in energy prices compared with the rest of the nation. Employment would be little affected or slightly benefited. Income transferred from natural gas consumers in the region would be recycled to the region through tax dollars paid by producers to the federal government, which would redistribute it, and through revenues from pension funds and from dividends of and investments in natural gas producers.[208]

Representative William E. Dannemeyer (R-Calif.) contends that differences in the amount of old gas under contract held by interstate pipelines as compared with holdings of such gas by intrastate pipelines discriminates against gas consumers served by intrastate pipelines.[209] Intrastate pipelines, in time, should be better able to compete with interstate pipelines for gas in the states that they serve after decontrol.

Some opponents of deregulation believe that deregulation of wellhead prices will benefit regions that produce natural gas, thus pitting consuming states against producing states. Projections by the American Gas Association for 1981 do indicate that residential consumers in two areas, the west south central and Pacific regions, areas that produce natural gas, would have lower percentage increases in their gas bills after four years of decontrol than would be true of other regions. Prices in other regions, however, are not predictable on this basis alone.[210]

Some regions are, of course, endowed with major coal resources and cheap hydroelectricity. Geography provides the southern states with an advantage over northern states in that they need less heat during the colder months.[211]

Efficiency Concerns

Opponents of deregulation contend that a regulatory regime is required to control gas prices. Pipelines are public utilities, and rate-of-return regulation best serves the public interest. Proponents of deregulation respond that market forces will promote efficiency and best serve the public interest.

Incentives for Efficiency. Advocates of deregulation contend, first, that the present pricing system with rolled-in average prices and automatic pass-through of purchased-gas prices, with the add-on of cost of transportation, does not provide either prompt or accurate market signals to either producers or consumers.[212] As long as consumer demand seems relatively inelastic, pipelines have little feel for the price at which consumers will switch to other fuels or reduce purchases through conservation.

This, and add-on pricing through the regulatory process, gives pipelines and distributing firms little incentive to strike the best bargain possible in negotiating supply contracts.[213] Most residential consumers are poorly informed as to the marginal cost of gas or as to probable future gas prices and supplies. Both industrial and residential consumers may make improvident investments in gas-burning equipment because of a lack of prompt and full information.[214] Producers and pipelines may not fully realize consumer resistance to the prices they have agreed upon until a gas glut develops.[215] In a deregulated market, it is claimed, wellhead prices would be established by a net-back system rather than by the traditional add-on system.[216] Scholars contend that effective energy policy requires a pricing system which signals to producers and importers the correct worth of natural gas to consumers and which lets consumers know what natural gas is costing society.[217]

Defenders of the current regulatory structure contend that pipelines enjoy economies of scale that require their regulation in the public interest. Distributing companies generally enjoy a monopoly because it would be inefficient to lay duplicative local distrubution systems for natural gas. Many participants in the natural gas markets—producers, pipelines, and numerous consumers, particularly industrial consumers and utilities—know the costs of competing fuels and the more knowledgeable ones follow regulatory proceedings and can determine pricing of natural gas. In sum, defenders of regulation contend that market signals are transmitted now and producers know the prices at which industrial users are likely to switch from gas to other fuels.

Allocative Efficiency. Economists argue that deregulation is desirable, for, in time, marginal cost pricing would substantially replace roll-in pricing and generate the efficiencies the current pricing system lacks. Demand would be measured by consumers' aggregate willingness to buy, and supply by producers' aggregate willingness to sell. Competitive establishment of an equilibrium price would improve allocative efficiency.

According to Richard J. Pierce, Jr.,

> Efficiency results because the value to society of the last unit purchased equals its cost of production exactly. If excess profits begin to develop, other producers enter the industry, supply expands, and excess profits disappear. A price ceiling upsets the perfectly competitive result.
>
> Producer regulation generates substantial transaction costs, including the cost of administering producer price controls and the cost of allocating gas administratively among consumers. Second, producer regulation creates regulatory uncertainty for downstream suppliers and consumers, who are unable to predict the extent of their future access to natural gas. This uncertainty inhibits investments that would be made in a stable market.[218]

Currently producers are deterred from producing additional quantities of price-regulated natural gas. They cannot afford to invest to produce added quantities, it is claimed, at regulated prices. The rewards for producing unregulated supplies of gas far exceed those for regulated supplies. Thus, some old gas will remain unused in the ground under regulation when a decontrol price would have brought it to market.[219]

Similarly, society is currently the loser when price controls on old gas and roll-in average pricing encourage the development of and sale of gas that is more expensive than the equilibrium price of competing energy sources, particularly fuel oil. The true cost of this gas, hidden by price averaging, would be disclosed in a market where supply and demand are free to set prices.[220] That system of pricing would prevent the uneconomic utilization of more costly natural gas.

To the extent, if any, that inclusion of price-controlled gas reduces the average price at the burner tip below the price that would be set by supply and demand in the fuel market, efficiency is sacrificed because consumers use too much gas in the present. Controlled prices also reduce incentives to explore for and produce natural gas and thus reduce supply and force some users to substitute expensive alternative fuels, an additional economic loss.[221]

Real economies are realized, it is claimed, when producers secure their gas supplies from sources in the order of their economic attractiveness, as would be true with deregulation.[222] Price decontrol, according to the CBO, permits prices rather than regulations to allocate supplies to end users. This produces more efficient allocation of existing supplies and encourages the production of additional supplies.[223] If prices of gas increase, this encourages fuel switching, thus freeing natural gas for use where its special characteristics are most valuable.[224]

Dismantling price controls and legislative constraints on the use of natural gas, it is claimed, would produce other efficiency gains. The NGPA is so complex that one major company estimated that its executives spend a third of their time coping with its regulations.[225] Deregulation would end this burden and eliminate the need for a large bureaucracy, legal fees, and court costs.[226] Deregulation would promote efficiency through conservation measures and the acquisition of fuel-efficient energy equipment. It would, in time, correct market disorders that plague the natural gas industry now.[227]

Advocates of regulation contend that regulation can be modified to deal with escalating prices. Competition from other fuels will force pipelines to observe prudent procurement practices and to minimize prices at the burner tip. The energy market across all fuels is competitive, and the goal of allocative efficiency is already being met in that context. This, it is claimed, is forcing more efficient practices within the gas industry without the need for legislative changes.

Reliance on market forces means gas prices will be heavily influenced by prices for fuel oil. World oil prices are volatile. Pipelines signing long-term supply contracts to ensure financing for construction could not, according to some defenders of regulation, afford to match short-term oil price swings. Buying on short-term supply contracts or on a spot market basis would make borrowing for pipeline systems construction difficult.

General Economic Impact

Some scholars feel that it is impossible to calculate the efficiency losses from the factors cited in the preceding section.[228] Scholars do, however, employ models to estimate the macroeconomic effects of public policy changes such as the deregulation of natural gas.[229] Although results depend greatly on the assumptions underlying these models and many will disagree with the projections, the following enumeration may be of some assistance in looking to the future. Short-run effects may be marked by some disorder. Long-run effects are cited as particularly important.

Inflation. The CBO notes that while energy prices would increase the consumer price index, reduced consumption of nonenergy goods and services would dampen this increase. Some producer prices would increase because of their higher energy costs, and this would indirectly increase the CPI. Wage rates indexed for inflation would produce a round of indirect wage increases that would affect the CPI.[230]

Milton Russell acknowledges that the price-level effect of changes in energy policy are important but claims the estimates of such changes can easily be overstated. Consumer price increases in energy do not match the magnitude of the price at the wellhead because other costs make up as much as half of the burner tip price. Energy costs are a relatively small part of the overall economy. The economy regularly sustains price fluctuations in other sectors that induce reallocation of productive resources and the restructuring of consumer demand patterns. He believes that energy should be viewed in this historical context.[231]

Economists Mark Ott and John Tatom estimated that with a gas price increase of 9.3 percent to 27.5 percent, the general level of prices would be about 1.1 percent to 2.2 percent lower than otherwise within about a year after decontrol. A worst-case scenario of a 50 percent increase in gas prices would temporarily raise the level of prices by only 0.4 percent within one year.[232]

Projections published in the *Morgan Guaranty Survey* based on a 20 percent price increase for residential consumers and a 30 percent increase for industrial consumers with decontrol showed that there would be a half percent increase in the inflation rate, but this rate would be lower in out-years. This study concluded that inflation expectations were less likely to be stimulated by a one-time jump in prices.[233]

The Energy Information Administration has published projections estimating, based on a CPI of 100 in 1967, that the 1986 CPI under continued natural gas price controls would stand at 331, under the NGPA at 332, and, with adoption of the administration's proposal, at 333. Comparable 1990 estimates are 464, 466, and 471.[234] Some policy makers have criticized the assumptions used in making these projections and contend that wellhead prices would not be as high as estimated.[235]

Some observers, as noted earlier, believe prices may not rise with deregulation in view of the current gas surplus and legislative provisions to deal with contract clauses, prudent purchasing by pipelines, and so forth.

Unemployment. There would not be a noticeable change in employment in the long run, according to one study, though increased gas prices could affect unemployment noticeably in the short run.[236] EIA projections show no difference in 1986 between unemployment under continued controls and the NGPA, and a one-tenth of 1 percent increase under the administration's proposal. All three models show the same unemployment rate in 1990.[237] There should be no change in employment rates, of course, if gas prices hold steady with deregulation. A 1981 Department of Energy study estimated that in the first three years after full decontrol the unemployment rate would rise 0.1 percent to 0.2 percent.[238]

Gross National Product. EIA projections of GNP show that deregulation under the administration's proposal may reduce GNP in 1986 by 0.2 percent, by 0.4 percent in 1988, and by 0.2 percent in 1990.[239] The CBO, however, states that, after an initial period of dislocation that would reduce national income, the economy would begin to adjust and pave the way to higher output and employment. Gas producers would recirculate income by spending increased revenues on new investment projects and additional workers. These changes would allow the economy to produce more output with fewer inputs and thus raise the national income by realizing potential efficiency gains.[240]

Effect on Oil and Gas Imports

Advocates of repeal of the Fuel Use Act and the deregulation of wellhead prices for natural gas contend that these steps are necessary to reduce our dependence on foreign oil. Increased gas supplies stimulated by the legislative changes sought, it is claimed, would make the purchase of substantial quantities of foreign oil unnecessary. Estimates of the effects of deregulation on oil imports vary. A 1981 DOE study estimated that immediate price decontrol would reduce imports by 440,000 barrels a day through 1985. Lower gas prices after that date would not stimulate additional gas production, as would higher initial gas prices, and oil imports would, by this estimate, increase by 190,000 barrels a day in 1990.[241]

A recent DOE analysis estimates that the administration's bill will reduce oil imports by 250,000 to 350,000 barrels a day more than would be true under the NGPA during the first year, and by 150,000 to 160,000 barrels a day the second year.[242]

Some observers believe gas supplies would not increase significantly under deregulation if prices remained level, and thus there would likely be no foreign oil displacement in the U.S. market. W. David Montgomery, an economist with Resources for the Future, argues that energy security would be enhanced by total deregulation, since this would give the gas industry flexibility to respond to future price changes in foreign oil in a way that would not be possible under the NGPA.[243]

Economist Richard Cooper argues that deregulation would not produce additional natural gas supplies to displace foreign oil. He feels that the amount of gas supply under a regime of continued regulation is underestimated.[244] Total deregulation without legislative action to modify indefinite escalator clauses, according to George H. Lawrence, president of the American Gas Association, would increase oil imports by 500,000 to 1 million barrels of oil a day in 1985 as compared with NGPA and escalator clause restrictions.[245]

Some scholars believe industrial users have a clear choice—risk dependence on foreign oil suppliers who are subject to political and cartel constraints or rely on domestic gas producers who are only subject to economic and competitive conditions.[246] The advantages of reduced dependence on foreign oil include the following. Lessened dependence on foreign oil reduces vulnerability to supply interruptions and their adverse consequences for national defense and our economic stability. Overreliance on foreign oil, which may greatly increase in price over time, will erode consumer purchasing power and business profits and lower demand and investment. Excessive reliance on foreign oil produces a net outflow of dollars. Absent offsetting U.S. exports to the foreign oil-producing countries, this will lead to a weakened dollar, reduced national income, and increased domestic inflation. Dependence on foreign oil also affects our relations with other nations, including Russia.[247]

U.S. oil imports in 1982 were roughly 11 percent of total free-world crude oil demand. Thus, reduced U.S. dependence on foreign oil would exert downward pressure on world oil prices to the advantage of all net importing countries.[248]

Advantages of reliance on foreign oil imports include, of course, the conservation of domestic fuel reserves for future emergencies. Also, foreign oil is a flexible source of energy large enough to cushion the vagaries of demand caused by the weather, strikes, delays in nuclear power plant construction, and so forth.[249]

Some pending bills would limit the import price or the import of high-priced foreign natural gas or LNG when cheaper natural gas is "shut in" in the United States.[250] Consumers in Massachusetts and California in particular are concerned with the pass-through of these higher gas costs that affect fuel prices in their states.[251]

The administration's bill omits provisions dealing directly with these imports, which amount to about 5 percent of U.S. natural gas consumption. The administration reportedly does not wish to disrupt diplomatic initiatives in this area by the adoption of legislative constraints.[252]

The American Gas Association claims that substituting 1 Tcf of gas from Mexico or Canada for other gas imports would improve the U.S. balance of trade by $1.3–3.5 billion and help stabilize OPEC oil prices.[253]

6

GAS TRANSPORTATION REFORM

Numerous proposals in the Congress would compel natural gas pipelines to transport gas owned by others. S.1715, the Senate committee bill, for example, would empower the FERC and state agencies to order inter- and intrastate natural gas pipelines, respectively, to use excess capacity to transport gas for other shippers. The House version would limit the amount of natural gas available for such transportation to those volumes freed from contractual commitments under the proposed legislation.

Gas pipelines are engaged principally in buying and selling gas on their own behalf and transporting that gas to distributing companies and industrial customers. Gas pipelines may engage in contract carriage for other owners of gas, but, currently, they are not required to do so. Common carriers in the conventional meaning of that term are obligated to carry goods for all who tender them. If they unreasonably refuse transportation, they may be held liable. Regulated common carriers must, of course, provide service at a reasonable price and on a nondiscriminatory basis.

Most of the proposals discussed in the preceding chapter would remove government restrictions on wellhead prices and permit the market to discipline those involved in the natural gas market. Proposals dealing with the transportation function of pipelines, however, generally call for additional regulation of pipelines. This chapter reviews the arguments for and against additional regulation that would require pipeline carriage of natural gas for other owners when the pipelines have the capacity to carry additional gas.

End users of natural gas usually do not take a constant, continuous supply. Rather, they take at peak periods or five days a week or daytime only. Home-heating requirements obviously are much greater in winter than summer. Nevertheless, production at the wellhead is continuous and fairly constant. Because of the size of a pipeline's system, it can balance peak demand with pipeline capacity and store substantial quantities of gas. By purchasing from numerous suppliers, pipelines reduce the risk of supply interruptions.

Some industry observers believe that the unregulated pipeline functions of middleman or "broker" and of storage are separable from the regulated function of carrier of natural gas. Nonetheless, others that might like to perform the brokerage function could have substantial difficulties. Brokers arranging only a few transactions a year may not enjoy the economies of operation that pipelines have. Direct purchase by end users may be impractical for small industrial consumers and small distribution systems. Few brokers could provide storage.

Several modes of transportation can be used for moving other petroleum products, and those products can be stored at any point in the distribution chain. Advocates of enforced pipeline carriage for gas that is not owned by the pipelines maintain that transportation and storage alternatives for natural gas are extremely limited. Philip R. O'Connor, chairman of the Illinois Commerce Commission, insists that greater competition at the wellhead will not be sufficient to ensure the degree of competition that would yield desired efficiency gains. The missing ingredient in his view is making it possible for producers on the one hand and industrial users and distributors on the other to deal directly with each other and secure delivery of the gas involved through pipelines when capacity permits. Customers of no other transportation medium are similarly handicapped.[254] In sum, compelled carriage would remove artificial barriers to competition.

Producers of natural gas who cannot obtain contracts for the sale of their natural gas to pipelines and whose gas is thus shut in believe they could find a market for their gas if only they were able to negotiate directly with end users and obtain transportation by pipeline when the pipeline has capacity to transport the gas.[255] Tussing argues that producers could sell the natural gas that the pipelines will not take at lower prices if they could deal directly with large consumers. Direct sales would help restrain prices and maintain exploration and development incentives.[256]

Commissioner O'Connor believes that all new gas production, all gas not under contract, and all gas released through the renegotiation of gas supply contracts should be carried for owners by pipelines when capacity permits. He believes that, in view of current market disorders, compelled carriage of natural gas could lead to reduction of gas prices of up to 20 percent at the burner tip.[257]

A study by the Illinois Commerce Commission concludes that converting pipelines to common carriers and modifying gas sales contracts consistent with this status are critical to achieving lower gas prices. The study estimates that common carriage added to the NGPA or the administration's bill could cut wellhead prices by 3 percent.[258] Advocates of compelled carriage claim that FERC rules for contract carriage are inadequate.[259] Compelled carriage is needed to permit the functioning of spot and futures markets that would establish prices based on real market conditions.[260] Knowledge of these prices would enable producers, pipelines, and consumers to obtain current market signals so they need not be surprised so often by unanticipated developments. Compelled carriage would help assure that producers in certain areas would not be left at the mercy of a single pipeline.

Pipelines that transport purchased gas can more easily match production and consumption than can direct dealing between producers and gas distribution companies or industrial users. Pipelines reduce the risk of service interruptions by pooling supplies of gas from a number of sources and then reselling the gas to distributing companies and industrial users. End users who buy direct from a single producer would not have these protections. Any benefits from increased competition would be more than offset, it is claimed, by the danger stemming from diffusion of responsibility among many competing companies, none of which may be permanently committed to the market.[261]

Some proposals would permit independent brokers to use a pipeline system to sell to the pipeline's own customers. A high volume of direct sales to industry is not likely, it is claimed, since these users would pay a high price for bypassing the lines of local distributing companies. Local distributing companies regulated as monopolies fear that industrial users will bypass their systems, thus forcing household consumers to pay a high part of the companies' fixed costs, which in turn may cause additional load loss.

Critics of these reform proposals claim that most gas-producing areas are served by two or more pipelines and gas is consumed in regions served by two or more pipelines. Thus, there is more competition than advocates of compelled carriage admit. The principal reason that pipelines discourage applicants for contract carriage, it is asserted, is that

FERC rules and procedures for contract carriage are cumbersome and greatly delay approval of these arrangements. The FERC has recently issued order 319, which should make it easier for producers and others to arrange contract carriage.

Pipeline firms must make a major investment in their facilities. Their pipelines running hundreds and thousands of miles are buried in the ground and cannot be readily moved as business conditions change. Thus, pipelines are entitled to perform their public utility functions in a manner that will enable them to retire their investments in equipment with a reasonable rate of return. Radical restructuring of the regulatory regime might seriously damage the viability of the firms that have done so much to make clean-burning, environmentally safe natural gas available to so many people.

If well-financed industrial users develop significantly better arrangements for gas than others, such as local distribution companies, inequities may develop. There is a limited supply of lower-priced natural gas. Residential consumers, served by local distribution companies, may be forced to pay even higher prices than they are charged now if their distributors are relatively less successful in acquiring low-cost supplies, either by direct negotiation or through the pipeline that serves their area.

The FERC could order pipelines to construct new facilities necessary to facilitate transportation under one bill. The transactions for which construction would be ordered might only last for a short period, such as two years. This would be an inadequate period of time for the pipeline to recover its capital costs for the facilities, such as extra pipe and compressor stations, which could cost millions of dollars. Then, because the "new" facilities might go unused after the compelled carriage transaction ends, the costs would be absorbed by all of the pipeline's other ratepayers. Some experts believe that compelled carriage may principally be used by large local gas distributing systems. Smaller systems would need the brokerage assistance of pipelines.[262]

Some critics object to the Bradley amendments, contending that industrial users would not be able to disconnect from local distribution companies that historically have served their plants. Commissioner O'Connor contends the Bradley amendments would deny local distributing companies access to lower-cost shut-in gas because of the minimum-bill and sole-supplier clauses in current pipeline/distributor contracts.[263] Some critics of the administration's bill feel it is too limited. Brokers could not participate under that bill unless they were able physically to take delivery of gas after completion of pipeline transportation, and few could do so.

Appendix A: Overview of the Natural Gas Policy Act of 1978

Section	*Description*	*Price Escalation Formula*	*Status as of Jan. 1, 1985*	*Average Prices per Mcf, May 1983*
Supply incentives				
102	New natural gas outside existing fields; new reservoirs; new outer-continental-shelf fields	Annual inflation plus real growth premium	Deregulated	3.54
103	New onshore wells within existing fields	Annual inflation	Deregulated	3.08
107	High-cost gas	Deregulated immediately	Deregulated	6.82
108	Stripper wells	Same as sec. 102	Regulated	3.89
Consumer protection				
104	Old interstate gas	Same as sec. 103	Regulated	1.41
106(a)	Renegotiated interstate contracts	Same as sec. 103	Regulated	1.16[a]
109	All other gas	Same as sec. 103	Regulated	2.45
Intrastate market				
105	Intrastate gas	Tied to new gas prices	Deregulated	2.28
106(b)	Renegotiated intrastate contracts	Same as sec. 103	Deregulated if contract price is greater than $1.00 per thousand cubic feet	1.16[a]

[a] Price is the average for interstate and intrastate contracts.

Sources: For provisions of the NGPA, see *Understanding Natural Gas Price Decontrol* (Washington, D.C.: Congressional Budget Office, April 1983), p. 6; for prices, see *Natural Gas Monthly*, Department of Energy, Energy Information Administration, June 1983, pp. 52–54.

Notes to Text

1. Stephen G. Breyer and Paul W. MacAvoy, "The Natural Gas Shortage and the Regulation of Natural Gas Producers," *Harvard Law Review,* vol. 86 (1973), p. 941.

2. U.S. Congress, Senate, *Hearings on Natural Gas Legislation Proposals,* 98th Cong., 1st sess., May 12, 1983, statement of Philip R. O'Connor, chairman, Illinois Commerce Commission.

3. *Information on Contracts between Natural Gas Producers and Pipeline Companies* (Washington, D.C.: General Accounting Office, 1983).

4. Paul Garfield and W. Lovejoy, *Public Utility Economics* (Englewood Cliffs, N.J.: Prentice-Hall, 1964), pp. 294–95.

5. Liquids and liquefiables consist primarily of propane and butane but include certain other liquids.

6. Correlative rights refer to the right of the owner of each property in a producing pool of oil or gas to produce his just and equitable share of the oil or gas without waste. Howard Williams and Charles Meyers, *Manual of Oil and Gas Terms,* 5th ed. (New York: Mathew Bender, 1981), pp. 150–51.

7. *Gas Facts* (Arlington, Va.: American Gas Association, 1981), p. 59, table 46.

8. See Mid-Louisiana Gas Co. v. FERC, *U.S. Law Week,* vol. 51 (1983), p. 5030.

9. *Natural Gas Monthly,* March 1983, p. 11.

10. Industrial consumers sometimes purchase gas directly from pipelines. In November 1982 the average price of gas in these direct sales was $3.98/Mcf. In 1980 the industrial sector was the largest consumer of natural gas, using 8.4 quadrillion Btu. The residential sector used 4.9 quads, electric utilities used 3.8 quads; commercial usage for buildings, street lighting, and so forth, amounted to 2.7 quads, and transportation consumed 0.65 quads. *Natural Gas: Use and Expenditures* (Washington, D.C.: Energy Information Administration, 1983), p. 8.

11. Robert Stobaugh and Daniel Yergin, eds., *Energy Future: The Report of the Harvard Business School Energy Project* (New York: Ballantine, 1980), pp. 69–70.

12. 15 U.S.C. 717-717w.

13. The FPC was superseded by the FERC, a five-member independent regulatory agency within the Department of Energy, in 1978. See 42 U.S.C. 7171-7177 and Executive Order 12,009, 3 C.F.R. 142.

14. Panhandle Eastern Pipe Line Co. v. Public Service Commission of Indiana, 332 U.S. 507, 516 (1947). The court stated:

> Three things and three only Congress drew within its own regulatory power, delegated by the [Natural Gas] Act to its agent, the Federal Power Commission. These were: (1) the transportation of natural gas in interstate commerce; (2) its sale in interstate commerce for resale; and (3) natural gas companies engaged in such transportation or sale.

15. 15 U.S.C. 717(b).

16. The Hepburn Act made oil pipelines common carriers in 1906 but exempted natural gas pipelines from common carrier status. 49 U.S.C. 1(1)(b). See The Pipeline Cases, 234 U.S. 548 (1914). Natural gas pipelines operating on the outer continental shelf are not denominated common carriers. They are required, however, to transport or purchase "without discrimination natural gas produced from said submerged lands in the vicinity of the pipeline in such proportionate amounts as [the commission may] determine to be reasonable." 43 U.S.C. 1334(e). Compare William Mogel, "Ratemaking for Oil Pipelines in the Outer Continental Shelf," *Tulsa Law Journal,* vol. 17 (1982), p. 469.

Common carrier status for gas pipelines was proposed as early as 1913, and the Senate passed a bill to this end. *Congressional Record,* vol. 50 (1913), pp. 5847, 5848–49. Industry opposed common carrier status in House hearings, and the House took no action on the proposal. U.S. Congress, House of Representatives, Committee on Interstate and Foreign Commerce, *Hearings on S. 3345,* 63d Cong., 2d sess., 1914. Common carrier status was proposed again during consideration of legislation for the regulation of gas pipelines in the mid-1930s. U.S. Congress, House of Representatives, Committee on Interstate and Foreign Commerce, *Hearings on H.R. 5423,* 74th Cong., 1st sess., 1935; U.S. Congress, House of Representatives, Committee on Interstate and Foreign Commerce, *Hearings on H.R. 11662,* 74th Cong., 2d sess., 1936. The Natural Gas Act adopted in 1938 at the culmination of these hearings did not include common carrier status for gas pipelines.

17. Phillips Petroleum Co. v. Wisconsin, 347 U.S. 672, 677 (1954), construing section 2(6) of the NGA, 15 U.S.C. 717a(6).

18. Ibid., p. 682.

19. See, for example, Phillips Petroleum Co., 24 FPC 537 (1960), affirmed, 303 F.2d 380 (D.C. Cir. 1963). The FPC originally employed the cost-of-service approach to proper pricing for sales by gas producers.

20. See Southern Louisiana Rate Cases v. FPC, 428 F.2d 407, 416 (5th Cir.), cert. denied, 400 U.S. 950 (1970).

21. Atlantic Refining Co. v. Public Service Commission of New York, 360 U.S. 378, 391 (1959).

22. The FPC announced in 1960 its identification of some twenty-three producing areas and its intention to initiate area rate proceedings to review prices. See FPC Statement of General Policy 61-6, 24 FPC 818, modified, 24 FPC 1107, modified, 25 FPC 595 (1961). See also William Mogel, "The Federal Power Commission's Authority to Set Area Rates by Rulemaking," *Seton Hall Law Review,* vol. 5 (1973), p. 31.

23. See Area Rate Proceeding, FPC Docket No. AR 61-1, "Order Instituting Area Rate Hearings, Consolidating Proceedings, and Prescribing Preliminary Procedures," 24 FPC 1121 (1960).

24. See Permian Basin Area Rate Proceeding, 24 FPC 159 (1965), affirmed in part and reversed in part, Skelly Oil Co. v. FPC, 375 F.2d 6 (10th Cir. 1967), affirmed, *In re* Permian Basin Area Rate Cases, 390 U.S. 747 (1968).

25. See, for example, Phillips Petroleum Co. v. FPC, 475 F.2d 842 (10th Cir. 1973), cert. denied, 414 U.S. 1146 (1974); Southern Louisiana Area Rate Cases, 428 F.2d 407 (5th Cir.), cert. denied, 400 U.S. 950 (1970).

26. Opinion 699, 51 FPC 2212, 2215 (1974), affirmed, Shell Oil Co. v. FPC, 520 F.2d 1061 (5th Cir. 1975), cert. denied, 426 U.S. 941 (1976); Opinion 770, 56 FPC 509 (1976), affirmed *sub nom.,* American Public Gas Association v. FPC, 567 F.2d 1016 (D.C. Cir.), cert. denied, 435 U.S. 907 (1977).

27. Shell Oil Co. v. FPC, 520 F.2d 1061, 1072 (5th Cir. 1975), cert. denied, 426 U.S. 941 (1976).

28. Compare Barry Commoner, "A Nearly Perfect Fuel," *New Yorker,* May 2, 1983, p. 66.

29. See generally Paul W. MacAvoy, "The Natural Gas Policy Act of 1978," *Natural Resources Journal,* vol. 19 (1979), p. 811; Robert P. Nordhaus, "Producer Regulation and the Natural Gas Policy Act of 1978, *Natural Resources Journal,* vol. 19 (1979), p. 829.

30. When a natural gas pipeline lacks adequate supplies of gas to fulfill its contractual obligations, it must curtail deliveries to certain customers. The United States first experienced chronic natural gas shortages in 1970. See FPC v. Louisiana Power & Light Co., 406 U.S. 621, 626, and note 230 (1972); William Mogel, "Food, Fuel, and Federal Curtailment Regulation," *Chicago-Kent Law Review,* vol. 56 (1980), p. 789.

31. Richard J. Pierce, Jr., "Natural Gas Regulation, Deregulation, and Contracts," *Virginia Law Review,* vol. 68 (1982), pp. 63, 86 note 94, 86–90.

32. 15 U.S.C. 3315(b)(3)(A), 3316(b) (regulating sale of gas under existing intrastate contracts); 15 U.S.C. 3315(b)(3)(D) (providing for uniform treatment of new natural gas whether interstate or intrastate).

33. The "dual market" involves sales to interstate and intrastate purchasers. The FPC's "just and reasonable" price regulation, after *Phillips,* applied only to gas sold in the interstate market. Producers could avoid FPC price regulation and sell at a higher price in the intrastate market. Federal price regulation thus created a bifurcated or dual market. Intrastate pipelines had adequate supplies of natural gas. The interstate gas supply shortage under a regulatory regime was a major reason why Congress adopted the NGPA. Stobaugh and Yergin, *Energy Future,* pp. 72–73.

34. Section 121 of the NGPA, which provides for the elimination of price controls for certain categories of natural gas, does not decontrol the price of gas defined under section 104. See 15 U.S.C. 3331, 3314.

35. Section 101 of NGPA, 15 U.S.C. 3311.

36. Section 107(c), (5)15 U.S.C. 3317(c)(5). Gas prices deregulated by the FERC as involving extraordinary risks or costs include that produced from "tight sands" or other "tight formations." See *Natural Gas Monthly* for December 1982 for the percentage of high-cost gas produced.

37. Pennzoil Co. v. FERC, 645 F.2d 360, 366 (5th Cir. 1981), cert. denied, 454 U.S. 1142 (1982); and see Howard Williams and Charles Meyers, *Oil and Gas Law* (New York: Mathew Bender, 1977), section 726.1.

38. Pennzoil Co. v. FERC, 645 F.2d 360, 365 note 1 (5th Cir. 1981), cert. denied, 454 U.S. 1142 (1982).

39. *In re* Permian Basin Area Rate Cases, 390 U.S. 747 (1968).

40. Ibid., p. 782 note 46,783.

41. Order 329, 36 FPC 925 (1966).

42. Pennzoil Co. v. FERC, 645 F.2d 360 (5th Cir. 1981), cert. denied, 454 U.S. 1142 (1982).

43. The FERC announced a new general policy December 16, 1982, regarding prepayments for natural gas under take-or-pay provisions of natural gas contracts and amendments to such contracts that become effective after December 23, 1982. Take or Pay Provisions in Gas Purchase Contracts, Statement of Policy, 47 Federal Register 57,268 (1982). Under this policy statement the FERC will apply a rebuttable presumption in general rate cases that prepayments to producers will not be given rate base treatment if made pursuant to take-or-pay requirements which exceed 75 percent of annual deliverability.

44. See "take-or-pay clause" in the glossary. Under current FERC regulations, all natural gas sales contracts executed after February 1, 1967, must allow the buyer to receive gas paid for but not taken for up to five years following payment, unless the contract term is shorter. 18 CFR 154.103.

45. See, for example, Peter A. Holmes, "The Implosion at Columbia Gas," *Fortune,* vol. 107 (May 2, 1983), p. 185.

46. Rates must be just and reasonable. Section 4(a) NGA, 15 U.S.C. 717c(a); FPC v. Hope Natural Gas Co., 320 U.S. 591 (1944).

47. See Order 452, 47 FPC 1049 (1972), amending 18 CFR 154.38(d)(4).

48. Section 601(c) NGPA, 15 U.S.C. 3431(c). See Columbia Gas Transmission Corp., 21 FERC par. 63,100 (1982), in which the administrative law judge denied pass-through of costs, in an initial decision, on the ground that certain costs amounted to an "abuse" under section 601(c)(2) of the NGPA.

49. On April 11, 1983, the National Energy Board reduced the border price for gas exported to the United States from \$4.94/Mcf to \$4.40/Mcf. As a result, several U.S. interstate pipelines have filed revisions in their tariffs to reflect a rate decrease. See, for example, Great Lakes Transmission Company, Docket RP83-34-001 (filed May 31, 1983), reducing gas purchase costs by \$47.5 million.

50. Pemex is a public corporation that is the only entity permitted to explore for, develop, and refine petroleum in Mexico. The Mexican government has asked for parity with the Canadian border

price. See, for example, DOE/ERA Opinion and Order 16-A, ERA Docket 79-31-NG (May 15, 1980).

51. See Distrigas Corp., 47 FPC 752 (1972); Trunkline LNG Co., Opinion 796, 58 FPC (1977). Purchases of Algerian gas by Trunkline are being reduced 40 percent for the next several months. "Energy: Trunkline to Take Lower Volumes of LNG, Asks FERC to Approve Modified Price," *Bureau of National Affairs Daily Report for Executives* (hereafter cited as *BNA DER*), July 14, 1983, p. A-13.

52. 15 U.S.C. 717b.

53. When Congress created the Department of Energy in 1977, it transferred most of the regulatory powers previously held by the FPC to DOE. 42 U.S.C. 7171, 7172. The secretary of the department delegated that authority to ERA and the FERC through Delegation Order 0204-54, issued November 1977. Responsibility for import/export authorization was one of the few responsibilities of the Federal Power Commission under the NGA that was not transferred to the FERC.

54. See West Virginia Public Service Commission v. United States DOE, 681 F.2d 847 (D.C. Cir. 1982).

55. Distrigas Corp. v. FPC, 495 F.2d 1057 (D.C. Cir.), cert. denied, 419 U.S. 834 (1974).

56. 15 U.S.C 717b.

57. Distrigas Corp. v. FPC, 495 F.2d 1057, 1065 (D.C. Cir.), cert. denied, 419 U.S. 834 (1974).

58. See Cia Mexicana de Gas v. FPC, 167 F.2d 804 (5th Cir. 1948).

59. See Trunkline LNG Co., 22 FERC pars. 61,245, 61,442 (February 28, 1983).

60. DOE/ERA Opinion and Order 50, Trunkline LNG Co., Docket 82-12-LNG (February 25, 1983).

61. See Panhandle Eastern Pipe Line Co. v. Public Service Commission of Indiana, 332 U.S. 507, 516 (1947).

62. Ibid., p. 516, note 46; and see FPC v. Louisiana Power & Light Co., 406 U.S. 621 (1972).

63. 15 U.S.C. 3371(a)(1). See also William Mogel and James G. White Jr., "An Assessment of Section 311 of the NGPA, *North Dakota Law Review,* vol. 58 (1982), p. 575.

64. 15 U.S.C. 3371(a)(2).

65. 15 U.S.C. 717f(c).

66. See 15 U.S.C. 3332.

67. A purchaser, under the administration's proposal, could not refuse gas if this would cause drainage or injure correlative rights.

68. 15 U.S.C. 3341-3348. See generally William Mogel and William R. Mapes, Jr., "An Assessment of Incremental Pricing under the NGPA," *Catholic University Law Review,* vol. 29 (1980), p. 763.

69. Bills introduced in the 98th Cong., 1st sess., include: S. 615 by Senator McClure; S. 689 by Senator John Heinz (R-Pa.); S. 996 by Senator Nancy L. Kassebaum (R-Kans.); S. 1049 by Senator Gary Hart (D-Colo.); S. 1119 by Senators Alan J. Dixon (D-Ill.) and Charles H. Percy (R-Ill.); H.R. 1752 by Representative Joseph P. Addabbo (D-N.Y.); H.R. 1760 by Representative Corcoran and six cosponsors; H.R. 2054 by Representative Berkley Bedell (D-Iowa); H.R. 2154 by Representative Richard A. Gephardt (D-Mo.) and seventy- three cosponsors; H.R. 2182 by Representative Patricia Schroeder (D-Colo.); H.R. 2499 by Representative Don Ritter (R-Pa.); H.R. 2508 by Representatives Jim Slattery (D-Kans.) and Daniel R. Coats (D-Ind.); and H.R. 2565 by Representative Corcoran and eleven cosponsors.

70. The Consumer Relief bills would amend section 110 of the NGPA, 15 U.S.C. 3320(a)(2), to permit recovery of certain production-related costs by the producer if necessary to permit a just and reasonable return.

71. An exception is provided for the first sale of casinghead or stripper well gas.

72. Consideration would include cash, in-kind compensation, or other remuneration. Under the Heinz bill, however, the FERC could provide an exception to the extent additional recovery is

necessary to permit the producer to earn a just and reasonable rate of return. The FERC would take into account all forms of consideration received by the producer, including free or low-cost transportation service for the producer's nongaseous hydrocarbons. The FERC's current policy was established in *Indicated Producers,* 22 FERC par. 61,621 (1983).

73. Excessive rates and charges subject to refund on and after October 1, 1979, include interest calculated at the average prime rate, as determined under FERC regulations. 18 CFR 154.67(c). The refund interest rate for the third quarter of 1983 is 10.5 percent on an annual basis.

74. According to section 3 of the Fair Marketing bill,

> The term "abuse" includes misrepresentation, imprudence on the part of the pipeline, failure by a pipeline to bargain at arms length with any producer, and the entering into of or operating pursuant to any contract by any pipeline with any producer if such contract materially prevents the pipeline from responding to changes in customer demand or other market forces.

75. 15 U.S.C. 3312(b). The maximum lawful price for this category of natural gas was $3.42/MMBtu in June 1983. 18 CFR 271.101.

76. The Fair Marketing bill (H.R. 2164) provides this right to a "pipeline."

77. The rate of return is the rate allowed to be earned or earned by a utility enterprise, generally calculated by dividing the net operating income by the rate base. *Regulation of the Gas Industry,* vol. 3.

78. *Energy Prices and Public Policy* (Washington, D.C.: Research and Policy Committee of the Committee for Economic Development, and the Conservation Foundation, 1982), pp. 54–69. United Gas Pipeline Co. v. Mobile Gas Corp., 350 U.S. 332, 344 (1956). Long-term commitments are essential to health of natural gas industry.

79. Representative William E. Dannemeyer, "Business Week Article 'Gas Pipelines,' " *Congressional Record,* vol. 128 (July 28, 1982), p. E 3555. For additional discussion, see Robert C. Means, "The Intrastate Pipelines and the Natural Gas Policy Act," in Edward J. Mitchell, ed., *The Deregulation of Natural Gas* (Washington, D.C.: American Enterprise Institute, 1983), pp. 71, 77.

80. Alan Murray, "Pressure from Consumers Pushes Congress into Action on Pricing of Natural Gas," *Congressional Quarterly,* vol. 41 (March 5, 1983), pp. 443, 444; *Energy Prices and Public Policy,* pp. 54–56.

81. Compare the choices cited in Catherine Good Abbott and Stephen A. Watson, "Pitfalls on the Road to Decontrol: Lessons from the Natural Gas Policy Act of 1978," and in Henry D. Jacoby and Arthur W. Wright, "The Gordian Knot of Natural Gas Prices," in Mitchell, *The Deregulation of Natural Gas,* pp. 66–68 and 141–44, respectively.

82. See Harry G. Broadman and W. David Montgomery, *Natural Gas Markets after Deregulation* (Washington, D.C.: Resources for the Future, 1983) for a discussion involving pipelines and distributors.

83. Cited in Senate Document 96-116, p. 440; and in Edward J. Mitchell, ed., *Perspectives on U.S. Energy Policy* (New York: Praeger Publishers, 1976), pp. 66–67.

84. "Natural Gas Decontrol: Issues and Prospects," *Morgan Guaranty Survey,* March 1982, pp. 7, 10.

85. *Understanding Natural Gas Price Decontrol* (Washington, D.C.: Congressional Budget Office, 1983), pp. 10, 26.

86. "Gas Consumers Say Pipelines Are Paying More Than Necessary for Supplies," *Wall Street Journal,* November 11, 1982.

87. Comments of Marlowe at the American Enterprise Institute conference on June 14, 1983, on "Implications for Natural Gas in a More Competitive Market" (hereafter cited as AEI conference).

88. Joseph Mulholland, *Economic Structure and Behavior in the Natural Gas Production Industry* (Washington, D.C.: Federal Trade Commission, 1979), p. 26.

89. I. C. Bupp and Frank Schuller, "Natural Gas: How to Slice a Shrinking Pie," in Stobaugh and Yergin, *Energy Future,* pp. 56, 72.

90. Lynn E. Browne and Richard F. Syron, "The Deregulation of Natural Gas: Its Potential Impact on New England," *New England Economic Review* (September/October 1976), pp. 45, 49.

91. Southern Louisiana Area Rate Cases, 428 F.2d 407, 416 n. 10 (5th Cir. 1970). See Senate Document 96-116, pp. 440–41, for the higher concentration in most manufacturing industries.

92. For the claim of increasing pipeline competition made over twenty years ago, see Paul MacAvoy, *Price Formation in Natural Gas Fields* (New Haven, Conn.: Yale University Press, 1962), pp. 243–50, 265. John H. Shenefield, when assistant attorney general for the Antitrust Division, argued that government regulation is incapable of anything more than broad adjustments. Such regulation has created more problems than it has solved in the energy field. He argues that we should give competition a chance. John H. Shenefield, "An Antitrust Perspective on Energy Regulation: Giving Competition a Chance," *Federal Bar News*, vol. 26 (December 1979), pp. 293, 295.

93. Senate Document 96-116, p. 441.

94. Ibid., p. 427.

95. *Understanding Natural Gas Price Decontrol*, p. 8.

96. See, for example, Sam H. Schurr, *Energy in America's Future: The Choices before Us* (Washington, D.C.: Resources for the Future, 1979), p. 235; Owen Phillips, *The Last Chance Energy Book* (Baltimore: Johns Hopkins University Press, 1979), p. 46; U.S. Congress, House of Representatives, Committee on Energy and Commerce, Subcommittee on Fossil and Synthetic Fuels, *Hearings on Natural Gas Outlook*, 97th Cong., 1st sess. (hereafter cited as *House Outlook Hearings)*, June 1 and 9, 1981, p. 118, attachments to statement of George H. Lawrence, president of American Gas Association.

97. "Energy: Oil Prices Seen Rising by 50 Percent by End of Century," *BNA DER*, July 19, 1983, pp. L-1, L-2.

98. *House Outlook Hearings*, p. 22, chart V attached to the statement of Albert H. Linden, Jr., acting administrator of the Energy Information Administration.

99. Ibid., p. 264, attachment to statement of C. M. Allen, petroleum geologist, on behalf of the Natural Gas Supply Association; Schurr, *Energy in America's Future*, p. 229. For various estimates of production in particular future years and potential supplies, see additional statements in *House Outlook Hearings*.

100. Stobaugh and Yergin, *Energy Future*, p. 50, citing U.S. Geological Survey estimates ranging from forty-four to a hundred years, Central Intelligence Agency estimates of from fifty to sixty years, and oil company estimates ranging from fifteen to thirty years; G. Henry M. Schuler, "Getting Government Out of Gas Regulation," *Wall Street Journal*, July 6, 1983 (estimates of half a century). Julian L. Simon, *The Ultimate Resource* (Princeton, N.J.: Princeton University Press, 1981), p. 107, cites estimates that are even more extreme, ranging from President Carter's estimate of ten years based on 1974 data to 1,000–2,500 years. One scholar cites an estimate that 97 percent of the potential natural gas fields in North America have not been discovered, and a *Wall Street Journal* estimate that as much as a thousand years' supply awaits discovery if prices are decontrolled. Barry P. Brownstein, "Natural Gas: The Case for Immediate Deregulation," *Policy Report*, vol. 4 (August 1982), pp. 1, 3.

101. *Energy: The Next Twenty Years* (Cambridge, Mass.: Ballinger, 1979), p. 71.

102. Paraphrased in Robert B. Helms, *Natural Gas Regulation: An Evaluation of FPC Price Controls* (Washington, D.C.: American Enterprise Institute, 1974), pp. 68–69; and in Mitchell, *Perspectives on Energy Policy*, p. 67.

103. Cited in Stobaugh and Yergin, *Energy Future*, p. 67.

104. Ibid., p. 75.

105. Ibid.

106. The results of these studies are discussed in Mitchell, *Perspectives on U.S. Energy Policy*, pp. 68–71; and in Browne and Syron, "The Deregulation of Natural Gas," pp. 51–52.

107. See, for example, Mitchell, *Perspectives on U.S. Energy Policy,* p. 75; Norman Metzger, *Energy: The Continuing Crisis* (New York: Thomas Y. Crowell Co., 1977), p. 89; Helms, *Natural Gas Regulation,* p. 71.

108. Pierce, "Natural Gas Regulation, Deregulation, and Contracts," pp. 90–92.

109. Cited in Clarence Brown, "On How We Got into This (Un)natural Gas Regulatory Mess," *Washington Times,* June 30, 1983.

110. *Supplemental Analysis of Natural Gas Consumer Regulatory Reform Legislation* (Washington, D.C.: Department of Energy, Office of Policy, Planning, and Analysis, 1983); *Analysis of the Administration's Natural Gas Decontrol Plan* (Washington, D.C.: General Accounting Office, 1983); *Comparison of Natural Gas Pricing Analyses* (Arlington, Va.: American Gas Association, 1983); C. S. Mathews, *Increase in United States "Old Gas" Reserves Due to Deregulation* (Houston: Shell Oil Company, 1983).

111. For a discussion of the types of contract clauses and how frequently they are employed, see U.S. Congress, Senate, Committee on Energy and Natural Resources, *Hearings on the Implementation of Title I of the Natural Gas Policy Act of 1978,* 97th Cong., 1st sess., November 5 and 6, 1981, pp. 332–33 (hereafter cited as *Senate Hearings on Title I*), statement of George H. Lawrence, president of the American Gas Association, and pp. 817–36, copy of report prepared by Energy Information Administration; Pierce, "Natural Gas Regulation, Deregulation, and Contracts," pp. 92–99. Helms, *Natural Gas Regulation,* pp. 12–16; *Understanding Natural Gas Price Decontrol,* pp. 7–9; *Natural Gas Producer/Purchaser Contracts* (Washington, D.C.: Energy Information Administration, 1982); Abbott and Watson, "Pitfalls on the Road to Decontrol," pp. 53, 62–69; *Information on Contracts between Natural Gas Producers and Pipelines* (Washington, D.C.: General Accounting Office, 1983).

112. *Guidance Needed on Use of Natural Gas Price Escalator Clauses* (Washington, D.C.: General Accounting Office, 1980), p. iii.

113. Milton Russell, "Overview of Policy Issues," in Mitchell, *Deregulation of Natural Gas,* pp. 3, 15.

114. See Abbott and Watson, "Pitfalls on the Road to Decontrol," pp. 62–66.

115. See, for example, the comments of the president of Tennessee Gas Transmission Company quoted in "Gas Pipeliners," *Business Week,* August 2, 1982, pp. 44, 48.

116. Abbott and Watson, "Pitfalls on the Road to Decontrol," p. 63.

117. See, for example, the views of Arlon R. Tussing and Connie C. Barlow cited in Russell, "Overview of Policy Issues," p. 16; the views of Merrill Lynch cited in Abbott and Watson, "Pitfalls on the Road to Decontrol," p. 63.

118. Abbott and Watson, "Pitfalls on the Road to Decontrol," p. 65.

119. Pierce, "Natural Gas Regulation, Deregulation, and Contracts," p. 94.

120. U.S. Congress, Senate, *Hearings on Natural Gas Legislation,* 98th Cong., 1st sess., March 28 and 29, 1983, part 2 (hereafter cited as *Senate Natural Gas Hearings*), p. 133, statement of Waidelich.

121. Ibid., p. 109, statement of Cleveland.

122. Comments of Marlowe at AEI conference.

123. Comments of Williams at AEI conference. Some 10 percent of the gas carried by Transco in June 1983 belonged to other shippers, and much of that gas had been shut-in gas that producers had been unable to sell to the pipelines. "Transco Energy: The Natural Gas Pipeliner That's Rewriting the Rules," *Business Week,* July 18, 1983, pp. 126, 127. For new purchasing and marketing practices of interstate pipelines, see Catherine Good Abbott, "Competition in Gas Markets: The Current Confusion," Paper presented at the North American meeting of the International Association of Energy Economists, June 9, 1983.

124. *Senate Natural Gas Hearings,* p. 32, statement of Marvin. Robert Woody says that producers cannot store natural gas and cannot eat it; they can only sell it. Thus, there will be

renegotiation of contracts without the need for legislation. Comments of Woody, in Mitchell, *The Deregulation of Natural Gas*, p. 48.

125. Comments of Representative Coats at AEI conference.

126. Comments of Tussing at AEI conference.

127. "Energy: Former FERC Official Says Shortages Possible with Wrong Policy Decisions," *BNA DER*, January 28, 1983, p. A-6, quoting the manager of corporate planning for MidCon Corporation.

128. Mulholland, *Economic Structure and Behavior in the Natural Gas Production Industry*, pp. 1–2.

129. Pierce, "Natural Gas Regulation, Deregulation, and Contracts," p. 75.

130. Paul Bennett and Deborah Kuenstner, "Natural Gas Controls and Decontrol," *Quarterly Review*, vol. 6 (Winter 1981–1982), p. 57; Senate Document 96-116, p. 439. Some scholars contend that failure to deregulate old gas would be interpreted by producers as a sign that Congress intends to retain some controls. There would thus be smaller increases in new supplies because of this uncertainty. Helms, *Natural Gas Deregulation*, p. 68.

131. Richard B. Mancke, *Providing for Energy* (New York: McGraw-Hill, 1977), p. 59.

132. "Natural Gas Decontrol: Issues and Prospects," p. 10; current disparities in price are noted by Clyde H. Greenert, of the energy and transportation policy department of Union Carbide Corporation, in comments at AEI conference, and by William K. Tell, Jr., senior vice president of Texaco, Inc., at that conference.

133. Bennett and Kuenstner, "Natural Gas Controls and Decontrol," pp. 57–58.

134. *Senate Natural Gas Hearings*, p. 302, statement of Lewis.

135. Comments of Hansen at AEI conference.

136. Comments of Benjamin Schlesinger of Booz, Allen, and Hamilton, and of Robert Woody, of Lane and Mittendorf, in Mitchell, *The Deregulation of Natural Gas*, pp. 38 and 160; comments of Tell at AEI conference. DOE refers to two kinds of price fly-up. Market-induced fly-up may be caused, for example, by a large increase in the price of competing fuel, in this case oil. Contract-induced fly-up would stem from the operation of various pricing and escalator clauses in gas supply contracts. *A Study of Alternatives to the Natural Gas Policy Act of 1978* (Washington, D.C.: U.S. Department of Energy, 1981).

137. Statements of Robert A. Hefner III, president and managing partner of GHK Industries, and Hubert Gentry, Jr., senior vice president and general counsel of Entex, Inc., and Rolland E. Kidder, executive assistant to the president of Envirogas, Inc., and Arlon R. Tussing, president of Arlon R. Tussing and Associates, at AEI conference.

138. Statements of Tell, Douglas W. Williams, senior vice president of Lone Star Gas Company, and W. David Montgomery, senior fellow at Resources for the Future, at AEI conference.

139. See remarks of Tell at AEI conference.

140. Ibid.

141. See remarks of Gentry at AEI conference. Greenert at AEI conference stated that it is absolutely necessary to eliminate the multitiered pricing system.

142. *Senate Natural Gas Hearings*, p. 14, statement of Boswell.

143. See remarks of Lay at AEI conference.

144. U.S. Congress, Joint Economic Committee, *Hearings on the Economics of Natural Gas Deregulation*, 98th Cong., 1st sess., April 15, 1983, p. 11, statement of MacAvoy.

145. Senate Document 96-116, p. 439. Advocates of decontrol argue that if decontrol produces higher natural gas prices, and if our projections of natural gas supplies are wrong and we have less gas than we thought, the conservation measures and fuel switching stimulated by higher prices will reduce gas consumption and give us time to develop alternative energy sources. In the meantime, limited supplies of gas would be used for the most highly valued uses. *Energy Prices and Public Policy*, p. 54.

146. "Natural Gas Decontrol: Issues and Prospects," p. 11. For an estimate that residential

consumer prices would be roughly 30 percent higher at the end of 1984 with full decontrol on January 1, 1983, see Bennett and Kuenstner, "Natural Gas Controls and Decontrol," p. 58.

147. See, for example, the comments of Representative John D. Dingell (D-Mich.), chairman of the House Energy and Commerce Committee, in "End Price Controls on Natural Gas?" *U.S. News & World Report,* vol. 91, November 30, 1981, p. 71; and the comments of Robert W. Stewart, chairman and chief executive officer of Michigan Consolidated Gas Company, at AEI conference.

148. *Understanding Natural Gas Price Decontrol,* p. 56.

149. Bennett and Kuenstner, "Natural Gas Controls and Decontrol," p. 56.

150. See Senate Report 94–191.

151. Comments of Hefner at AEI conference.

152. Ibid.

153. Comments of Stewart at AEI conference.

154. Comments of Gentry at AEI conference.

155. "Gas Users Turn to Congress for Relief as Prices Soar," *Congressional Quarterly,* vol. 40, October 30, 1982, pp. 2757, 2760. A recent DOE study concluded that the administration's bill would produce a reduction in wellhead and retail prices ranging from fifteen to thirty-five cents per Mcf below prices under the NGPA until 1987. If reserve additions were 10 Tcf, twice the amount assumed in this study, prices under the administration's bill would be lower than those under the NGPA until 1995. *Supplemental Analysis of Natural Gas Consumer Regulatory Reform Legislation* (Washington, D.C.: Department of Energy, Office of Policy, Planning, and Analysis, 1983).

156. Comments of Marlowe at AEI conference.

157. Comments of Williams at AEI conference.

158. See comments of Marlowe at AEI conference.

159. Art Pine, "Energy: The U.S. Has a National Policy and It's Working," reproduced in Senate Document 96-116, pp. 32, 38–39.

160. MacAvoy, *Price Formation in Natural Gas Fields,* p. 259; Neil de Marchi, "The Ford Administration: Energy as a Political Good," in Craufurd D. Goodwin, ed., *Energy Policy in Perspective: Today's Problems, Yesterday's Solutions* (Washington, D.C.: Brookings Institution, 1981), pp. 475, 509.

161. Comments of Lay at AEI conference.

162. Comments of Tell at AEI conference.

163. Comments of Greenert at AEI conference.

164. Jack Elam, spokesman of Process Gas Consumers Group, quoted in Lawrence Mosher, "Rising Natural Gas Prices—A Hard Lesson in Regulatory Miscalculations," *National Journal,* vol. 15 (January 1, 1983), pp. 24, 25. Clarence J. Brown cites experience with oil price controls as analogous. Oil prices climbed 912 percent during the control period from 1973 to 1980. U.S. oil production dropped 25 percent, and U.S. dependence on imports rose 50 percent. That, Brown claims, is what regulation did for us. Representative Clarence J. Brown, "Natural Gas Policy," *Congressional Record,* vol. 128, December 17, 1982, pp. E 5302, E 5303.

165. *Senate Natural Gas Hearings,* p. 17. statement of Kent S. Harrell, president of Harrell Energy Company.

166. Compare ibid., pp. 29–30, statement of Marvin.

167. Ibid., p. 30.

168. *Senate Natural Gas Hearings,* p. 119, statement of Waidelich, who argues that renegotiation of contracts will create backward pressure on new gas prices. Increased competition in today's buyers' market will also exert downward pressure on prices.

169. Ibid., pp. 6 and 15, statement of Cantrell.

170. Comments of Tussing at AEI conference.

171. Comments of Hefner at AEI conference.

172. Comments of Michael Canes in Mitchell, *The Deregulation of Natural Gas,* p. 149.

173. Comments of Tussing at AEI conference.

174. Comments of Tell at AEI conference.

175. Ibid.

176. *Washington Times* and other articles quoted in Representative Jack Fields, "The Time to Act Is Now," *Congressional Record,* vol. 129, May 11, 1983, pp. E 2186, E 2187; Pierce, "Natural Gas Deregulation and Contracts," p. 97; "Natural Gas Decontrol: Issues and Prospects," pp. 7–8; Bennett and Kuenstner, "Natural Gas Controls and Decontrol," pp. 56–57; comments of Rolland E. Kidder, executive assistant to the president of Envirogas, Inc., at AEI conference.

177. Hans H. Landsberg and Joseph M. Dukert, *High Energy Costs: Uneven, Unfair, Unavoidable?* (Washington, D.C.: Resources for the Future, 1981), p. 3.

178. Jacoby and Wright, "The Gordian Knot of Natural Gas Prices," p. 126.

179. See comments of Representative Coats at AEI conference.

180. See *Energy: The Next Twenty Years,* p. 50; *Energy Prices and Public Policy,* p. 59; *Understanding Natural Gas Price Decontrol,* pp. 10–11.

181. *Energy: The Next Twenty Years,* p. 49; Bennett and Kuenstner, "Natural Gas Controls and Decontrol," p. 57. For the high share of gas used for industrial purposes in five states, each with supplies of natural gas, see *Senate Hearings on Title I,* pp. 732–33, figures 9 and 10 attached to the supplemental comments of Donald R. Wills, president of Louisiana Resources Company, and pp. 423–25, testimony of Wills; *The Natural Gas Market through 1990* (Washington, D.C.: Energy Information Administration, 1983), p. 31.

182. Pierce, "Natural Gas Regulation, Deregulation, and Contracts," p. 73.

183. Comments of Tell at AEI conference.

184. *The Natural Gas Market through 1990,* p. 31.

185. Ibid., p. 11.

186. "Energy: Gas Producers' Treatment of Industrial Users Examined," *BNA DER* June 14, 1983, p. A-8.

187. *Understanding Natural Gas Price Decontrol,* p. 44; Landsberg and Dukert, *High Energy Costs,* pp. 30–31; *House Outlook Hearings,* p. 153, exhibits 1 and 3 to attachment 8 to statement of Lawrence.

188. Browne and Syron, "The Deregulation of Natural Gas," p. 62; and see comments of Hansen at AEI conference.

189. *House Outlook Hearings,* pp. 147–67, paper of American Gas Association entitled "Cost of Immediate Total Wellhead Price Decontrol of Natural Gas to Low Income and Disadvantaged Groups."

190. Pierce, "Natural Gas Regulation, Deregulation, and Contracts," p. 97.

191. See, for example, the 1981 American Gas Association projection for oil imports in Alexander Stuart, "The Blazing Battle to Free Natural Gas," *Fortune,* vol. 104 (October 19, 1981), pp. 152, 158.

192. Comments of Greenert at AEI conference.

193. Bennett and Kuenstner," Natural Gas Controls and Decontrol," p. 59; and see comments of Hansen, at AEI conference, on experience in Iowa.

194. Mitchell, *Perspectives on U.S. Energy Policy,* pp. 164–65. Compare Mancke, *Providing for Energy,* p. 81, as to one class of consumer benefiting at the expense of another with changes in fuel policy. Higher prices and fuel switching by some consumers would make gas available to those for whom the alternatives are worse. Milton Russell, *Energy* (Washington, D.C.: Resources for the Future, 1977), p. 325. Some customers are not nearly as concerned with the price of natural gas as with its availability. See Patricia E. Starratt, *The Natural Gas Shortage and the Congress* (Washington, D.C.: American Enterprise Institute, 1974), p. 58.

195. Quoted in Mancke, *Providing for Energy,* pp. 81–82. Landsberg and Dukert acknowledge the need for better statistics on the income distribution effects of energy prices. Landsberg and Dukert, *High Energy Costs,* p. 83.

196. Jack Stockfisch, "The Income Distribution Effects of a Natural Gas Price Increase," *Con-*

temporary Policy Issues, no. 1 (October 1982), p. 9. Stockfisch found that twice as many homeowners as renters use gas as a principal heating fuel. Thus, those families would bear a greater proportion of the burden of price increases. Half of the renter-occupied units heated with gas are located in the north central and northeast regions, and rents in those regions have grown more slowly in the 1970s than was true of rents in the South and West. Similarly, indirect consumption through purchase of products containing ingredients derived from natural gas or manufactured with the use of gas as a fuel is greater by families with higher incomes.

197. Ronald H. Schmidt, "Effects of Natural Gas Deregulation on the Distribution of Income," *Economic Review,* Federal Reserve Bank of Dallas, May 1983, p. 1; and see Glenn C. Loury, *Analysis of the Efficiency and Inflationary Impact of the Decontrol of Natural Gas Prices* (Washington, D.C.: Natural Gas Supply Association, April 1981); *A Study of Alternatives to the Natural Gas Policy Act of 1978* (Washington, D.C.: U.S. Department of Energy, Office of Policy, Planning, and Analysis, November 1981).

198. Robert S. Pindyck, "The Natural Gas Industry," *Current History* (May/June 1978), p. 229. Pindyck suggests that the food stamp program could be expanded by increasing stamp allotments and allowing the stamps to be applied to heating bills or that portion of rent allocated to fuels.

199. Russell, *Energy,* p. 322.

200. For the substantial disadvantage of pipelines with lesser endowments of old regulated gas in a partial decontrol situation, see Russell, "Overview of Policy Issues," pp. 17–19.

201. That natural gas pipelines are considered good investments even with the uncertainties involved with discussion of decontrol of gas prices, see "Gas Pipelines Become Tempting Targets," *Business Week,* June 20, 1983, p. 27. See *Information on Contracts between Natural Gas Producers and Pipeline Companies,* pp. 44–53, for a discussion of the potential price fly-up on deregulation and its impact on pipelines.

202. Murray, "Pressure from Consumers Pushes Congress into Action on Pricing of Natural Gas," p. 447.

203. Murray, "Gas Users Turn to Congress for Relief as Prices Soar," p. 2759; "Energy: Large, Small Producers Tell House Full Decontrol of All Gas Will Add to Reserves," *BNA DER,* September 24, 1982, p. A-6.

204. Milton R. Benjamin, "Gas Decontrol Would Fatten Big Oil Firms," *Washington Post,* July 20, 1983. The news article quotes the EIA administrator that the methodology of the report was faulty and that EIA is revising its conclusions.

205. Browne and Syron, "The Deregulation of Natural Gas," pp. 56–58.

206. Ibid., pp. 60-61.

207. Ibid., p. 63.

208. Robert A. Leone, "Impact of Higher Natural Gas Prices on the Northeast Economy," *Contemporary Policy Issues* (October 1982), pp. 1–8.

209. Dannemeyer, "Business Week Article: 'Gas Pipeliners,' " p. E 3555.

210. See *House Supply Hearings,* p. 164, appendix D to attachment 8 to the statement of Lawrence. For separate projections of gas prices for residential and industrial users under four alternatives, see *The Natural Gas Market through 1990,* pp. 59-60. All regions would reduce consumption of natural gas under projections for the future effect of the NGPA. Ibid., p. 47.

211. Senate Document 96-116, p. 442.

212. *Energy Prices and Public Policy,* p. 65.

213. *Understanding Natural Gas Price Decontrol,* p. 20; "Gas Pipeliners," p. 48, quoting Joseph P. Thomas, executive vice president of Peoples Gas Light and Coke Company of Chicago.

214. *Energy: The Next Twenty Years,* pp. 49, 149, n. 31; comments of Tell at AEI conference.

215. Compare comments of Gentry at AEI conference that the gas industry had no idea what the price of its product was.

216. Comments of Lay at AEI conference.

217. Schurr, *Energy in America's Future,* p. 439.

218. Pierce, "Natural Gas Regulation, Deregulation, and Contracts," p. 71.

219. Russell, "Overview of Policy Issues," pp. 23–24; Starratt, *The Natural Gas Shortage and the Congress,* p. 34; *Understanding Price Decontrol,* p. 22.

220. Bennett and Kuenstner, "Natural Gas Controls and Decontrol," p. 57; Russell, "Overview of Policy Issues," p. 24. Roll-in pricing is said to tax old gas to subsidize new gas. *Energy: The Next Twenty Years,* p. 49.

221. Pierce, "Natural Gas Regulation, Deregulation, and Contracts," pp. 71–72; Milton Russell, quoted in Starratt, *The Natural Gas Shortage and the Congress,* pp. 55–56; and see the high-cost sources, also cited in Starratt's book, p. 60.

222. *Energy Prices and Public Policy,* p. 65; comments of Kidder at AEI conference; "Natural Gas Decontrol: Issues and Prospects," p. 11.

223. *Understanding Natural Gas Decontrol,* p. 22; and see Schurr, *Energy in America's Future,* p. 442 (individual choices would replace many of the uncertain decision-models, questionable data, and imperfect knowledge on which government bases its allocative decisions).

224. Russell, *Energy,* p. 325.

225. "Natural Gas Decontrol: Issues and Prospects," p. 7.

226. Senate Document 96-116, p. 436.

227. *Senate Natural Gas Hearings,* p. 14, statement of Boswell; ibid., p. 37, statement of George W. Haney, chairman, Oklahoma Fertilizer Manufacturers Association.

228. See, for example, Russell, "Overview of Policy Issues," p. 24; *The Natural Gas Market through 1990,* p. 63.

229. For a discussion of macroeconomic models to analyze natural gas markets, see *Macroeconomic Forecasts of the Consequences of Natural Gas Deregulation* (Washington, D.C.: American Petroleum Institute, 1982). For a discussion of different analyses of natural gas deregulation, see a report prepared for the Chemical Manufacturers Association by Foster Associates, Inc., February 16, 1982, entitled "A Comparison and Appraisal of Ten Natural Gas Deregulation Studies."

230. *Understanding Natural Gas Decontrol,* pp. 30, 38.

231. Russell, *Energy,* pp. 321–22.

232. Mark Ott and John Tatom, "Are There Adverse Inflation Effects Associated with Natural Gas Decontrol?" *Contemporary Policy Issues* (October 1982), pp. 27–46. Ott and Tatom contend gas price decontrol would reduce demand for OPEC oil and therefore the world price of oil. They expect a 9.3 percent to 27.5 percent rise in gas prices to be matched by a 12 percent to 25 percent decline in overall energy prices.

233. "Natural Gas Decontrol: Issues and Prospects," pp. 11, 14.

234. *The Natural Gas Market through 1990,* pp. 64–65.

235. See letter from Representative Thomas Corcoran (R-Ill.) to the administrator of the EIA, May 24, 1983. The EIA acknowledged in its report that its analysis might overstate the negative impact of higher prices under decontrol. *The Natural Gas Market through 1990,* p. 63.

236. Schurr, *Energy in America's Future,* p. 449; compare William W. Hogan, "Energy and Economic Growth," in *Energy Conservation and Public Policy* (Englewood Cliffs, N.J.: Prentice-Hall, Inc., 1979), p. 11; Russell, *Energy,* pp. 321–22.

237. *The Natural Gas Market through 1990,* pp. 64–65.

238. *A Study of Alternatives to the Natural Gas Policy Act of 1978,* p. 29. Estimates of the effect of phased decontrol range from 0 percent to 0.2 percent. The study cautions that the macroeconomic models probably overstate the adverse impact of decontrol.

239. *The Natural Gas Market through 1990,* pp. 64–65.

240. *Understanding Natural Gas Price Decontrol,* pp. xiv, 39.

241. *A Study of Alternatives to the Natural Gas Policy Act of 1978.*

242. *Supplemental Analysis of Natural Gas Consumer Regulatory Reform Legislation.*

243. Comments of Montgomery in *Brookings Papers on Economic Activity,* no. 2 (Washington, D.C.: Brookings Institution, 1982), p. 392.

244. Richard Cooper, "A Note on Deregulation of Natural Gas Prices," *Brookings Papers on Economic Activity,* no. 2 (Washington, D.C.: Brookings Institution, 1982), pp. 371, 383.

245. U.S. Congress, Joint Economic Committee, Subcommittee on International Trade, Finance, and Security Economics, *Hearings on Energy Security,* 97th Cong., 2d sess., February 18, 1982, pp. 143–44, statement of Lawrence. Lawrence contends that total deregulation with activation of escalator clauses would produce a 2.5 Tcf loss in gas supply that would require the substitution of oil and coal.

246. Schuler, "Getting the Government Out of Gas Regulation."

247. The *World Oil Market in the 1980s: Implications for the United States* (Washington, D.C.: General Accounting Office, 1980), pp. x–xiv, 39–47, 53–65; *Understanding Natural Gas Price Decontrol,* p. 36 (reduced demand for foreign fuel can improve the balance of trade and increase national income); Schurr, *Energy in America's Future,* p. 449 (higher prices paid abroad reduce total goods and services available to U.S. citizens—a result that does not follow from increases in domestic energy prices).

248. See *Washington Times* editorial quoted in Fields, "The Time to Act Is Now," pp. E 2186, E 2187.

249. Thomas C. Schelling, *Thinking Through the Energy Problems* (New York: Committee for Economic Development, 1979), p. 27.

250. Mosher, "Rising Natural Gas Prices—A Hard Lesson in Regulatory Miscalculation," p. 25; and see *Senate Natural Gas Hearings,* p. 11, statement of Senator Don Nickles (R-Okla.).

251. Murray, "Pressure from Consumers Pushes Congress into Action on Pricing of Natural Gas," p. 446.

252. Ibid., pp. 446–47. For special problems with imports and exports of LNG, see "Gas Pipeliners," p. 46; and Metzger, *Energy: The Continuing Crisis,* p. 91. ERA has scheduled hearings on gas import policy for September 8, 1983.

253. *House Supply Hearings,* pp. 93, 119–34.

254. Comments of Philip R. O'Connor, chairman of the Illinois Commerce Commission, at AEI conference. Some advocates of gas pipeline reform recommend changes that would require that pipelines separate their brokerage and storage functions from their transportation function at least to the extent of separating charges for these services and accepting different regulatory regimes. See, for example, comments of Lay at AEI conference.

255. See, for example, comments of Marvin at AEI conference.

256. Comments of Tussing at AEI conference. Clarence J. Brown points out that at least one pipeline, Transco, has negotiated as go-between for its customers and producers seeking willing buyer and willing seller prices. It was found that the average price that was established was lower than current federally regulated ceiling prices. Brown, "On How We Got into This (Un)natural Gas Regulatory Mess."

257. Comments of O'Connor at AEI conference.

258. "Decontrol, Especially Coupled with Common Carriage, Is a Better Legislative Solution," *Inside FERC,* July 18, 1983, p. 7.

259. Comments of Richard E. Storat, manager of corporate energy affairs for Bethlehem Steel Corporation, at AEI conference.

260. Comments of O'Connor and Kidder at AEI conference.

261. See Deborah A. Cohn and Robert C. Means, *Contract Carriage of Natural Gas: A Preliminary Analysis of the Issues,* preliminary draft (Washington, D.C.: Federal Energy Regulatory Commission, 1983), pp. 7, 13.

262. Mitchell, ed., *The Deregulation of Natural Gas,* p. 160.

263. Comments of Commissioner O'Connor at AEI conference.

Glossary

Area Rate Clause. A price escalation clause in a gas sales contract that authorizes an increase in the contract price whenever regulatory authority approves increases in area rates—that is, the applicable just and reasonable wellhead price for the category of natural gas involved.

Bcf. Billion cubic feet.

Casinghead Gas. Natural gas obtained in the process of producing oil from an oil well.

Common Carrier. A carrier (for example, a railroad or oil pipeline) obliged by law to accept products for transportation without discrimination among shippers. Government authorities may regulate rates and require the building or acquisition of additional capacity to serve shippers.

Contract Carrier. A carrier that is obligated to accept products for transport without discrimination among shippers but who has no obligation to build or acquire additional capacity to handle additional volume.

Drainage. Drainage from under a property is the migration of oil or gas in a reservoir owing to pressure reduction caused by extraction of oil or gas at another point in the reservoir. The owner of oil and gas rights in a particular property may suffer loss because of drainage to a producing well on other property nearby.

EIA. Energy Information Administration within the Department of Energy.

ERA. Energy Regulatory Administration within the Department of Energy.

Favored-Nation Clause. An indefinite price escalator clause in a gas sales contract. See below. A two-party favored-nation clause obligates the purchaser to pay the producer at the highest rate the purchaser pays to any other producer in a specified producing area, such as in the same field. A three-party favored-nation clause enables the producer to collect for the delivery of gas at the highest price paid by any producer in a specified producing area.

FERC. Federal Energy Regulatory Commission, a regulatory commission within the Department of Energy whose jurisdiction includes activities subject to the Natural Gas Act and Natural Gas Policy Act.

Fly-up. A sudden increase in natural gas prices that may occur with the removal of price controls.

FPC. Federal Power Commission, an independent regulatory commission that has been superseded by the FERC.

Hinshaw Pipeline. A pipeline that is exempted from FERC jurisdiction by section 1(c) of the Natural Gas Act, provided all natural gas is received at or within the boundary of a state, the gas is all consumed within that state, and the rates, services, and facilities of the pipeline are subject to regulation by state authorities.

Incremental Pricing. Under the NGPA, a pattern of pricing that places a proportionately larger share of the cost of natural gas on industrial users.

Indefinite Price Escalator Clause. A provision in a gas sales contract that provides for the establishment or adjustment of the price for gas delivered under the contract by reference to prices for natural gas, crude oil, or refined petroleum products set by other contracts, regulatory authorities, or the market as the case may be. Prices under these clauses are "indefinite" because they are not known until fixed by happenings outside the control of the contracting parties.

LNG. Liquefied natural gas. Natural gas in its gaseous state is chilled until it liquefies so that it can be shipped in liquefied state.

MARKET OUT. An option provided in a buyer protection clause in many gas sales contracts that permits the purchaser of natural gas to cancel its obligation to take a certain volume of gas or some portion of that volume or, in some contracts, to pay a lower price for certain gas.

MCF. Thousand cubic feet.

MINIMUM-BILL REQUIREMENT. A clause in a gas sales contract that specifies that the price of or rate for the natural gas under contract shall not be less than a specified amount.

MMBTU. Million British thermal units.

MONOPSONY. A market situation in which there is one buyer and many sellers.

NGA. The Natural Gas Act of 1938.

NGPA. The Natural Gas Policy Act of 1978.

NEW GAS. Under the NGPA, new gas is gas produced from reservoirs from which gas was not produced in commercial quantities before April 20, 1977.

OFF-SYSTEM SALES. Sales by a gas pipeline to a customer outside its usual market area.

OLD GAS. Under the NGPA, old gas is gas dedicated to interstate commerce before enactment of the NGPA on November 9, 1978.

PGA. Purchased-gas adjustment. Pipelines under the NGPA may adjust regulated rates for the price of natural gas sold by them to reflect increases in their costs of purchased gas. Approvals of these rate adjustments, usually effected semiannually, have been automatic.

PRICE REDETERMINATION CLAUSE. A renegotiation, redetermination, or "better-market" clause that provides that at stated intervals—such as two years, three years, or five years—the parties will renegotiate or redetermine the contract price based on the prices then being paid in the field. If the purchaser and seller cannot agree, the price is fixed by arbitration. Arbitrators are frequently bound to take the average of the two or three highest prices being paid by pipelines in the field or area.

STRIPPER WELL GAS. Under the NGPA, gas not associated with the production of oil and produced from a well at a rate that does not exceed an average of 60 Mcf per production day during the preceding ninety-day production period.

TAKE-OR-PAY CLAUSE. Gas purchase contract clause obligating the purchaser to take a certain volume of natural gas and to pay for that volume even if it is not taken, usually a percentage of a total agreed volume that may be claimed under the contract. The take-or-pay obligation may apply to minimum daily quantities, to minimum annual quantities, or to minimum contract quantities. Most gas purchase contracts with take-or-pay clauses allow "make-up"—that is, the purchaser may obtain delivery of quantities paid for but not taken as scheduled within a specified time such as five years.

TCF. Trillion cubic feet.

TIGHT-SANDS GAS. Gas produced from geologic strata that hold gas too tightly for conventional extraction processes to bring it to the surface at economic rates without special stimulation.

Warranty Contracts. A contract for the first sale of gas that provides the purchaser an unconditional guarantee that all of the volume of natural gas agreed to be delivered will be delivered regardless of the source of the gas.

Wellhead. The point at which natural gas is transferred from the gas well to a pipeline or other non-well facility. Wellhead price is the price paid to producers.

The American Enterprise Institute for Public Policy Research, established in 1943, is a publicly supported, nonpartisan, research and educational organization. Its purpose is to assist policy makers, scholars, businessmen, the press, and the public by providing objective analysis of national and international issues. Views expressed in the institute's publications are those of the authors and do not necessarily reflect the views of the staff, advisory panels, officers, or trustees of AEI.

AEI Legislative Analyses

Reauthorization of the Clean Water Act
Review: 1982 Session of the Congress
The Clean Air Act: Proposals for Revisions
Proposed Revisions of the Foreign Corrupt Practices Act
Proposals for a Subminimum Wage for Youth
Proposals to Modify the Taxation of U.S. Citizens Working Abroad
Proposals to Regulate State and Local Taxation of Multijurisdictional Corporations
Recent Proposals to Restrict Conglomerate Mergers
The Patent Term Restoration Act of 1981
Review: 1980 Session of the Congress
Compulsory Health Insurance Proposals
Major 1980 Tax Cut Proposals
Youth Employment Legislation: The Youth Act of 1980
Military Retirement: The Administration's Plan and Related Proposals
The Administration's Plan to Reauthorize Revenue Sharing
The Capital Cost Recovery Act Proposal
Review: 1981 Session of the Congress